Planet Mars

Story of Another World

François Forget, François Costard
and Philippe Lognonné

Planet Mars

Story of Another World

Published in association with
Praxis Publishing
Chichester, UK

Dr François Forget
Research Scientist
Centre National de la Recherche Scientifique (CNRS)
Institut Pierre Simon Laplace
Université Paris-6
Paris
France

Dr François Costard
Senior Scientist
Centre National de la Recherche Scientifique (CNRS), IDES
Université Paris-Sud 11
Orsay
France

Professor Philippe Lognonné
Université Paris-7
Institut de Physique du Globe de Paris, CNRS
Saint Maur des Fossés
France

Original French edition: *La Planète Mars*
Published by © Éditions Belin 2006
Ouvrage publié avec le concours du Ministère français chargé de la culture – Centre national du livre
This work has been published with the help of the French Ministère de la Culture – Centre National du Livre

Translator: Bob Mizon, 38 The Vineries, Colehill, Wimborne, Dorset, UK

SPRINGER–PRAXIS BOOKS IN POPULAR ASTRONOMY
SUBJECT *ADVISORY EDITOR*: John Mason B.Sc., M.Sc., Ph.D.

ISBN: 978-0-387-48925-4 Springer Berlin Heidelberg New York

Springer is a part of Springer Science + Business Media (*springer.com*)

Library of Congress Control Number: 2007931635

Cover design: Jim Wilkie
Typesetting: BookEns Ltd, Royston, Herts., UK

Printed in Germany on acid-free paper

Table of Contents

PART 3
THE SLOW METAMORPHOSIS

PART 4
CLIMATES AND STORMS

PART 5
EXPLORING MARS
1650–2050

Acknowledgements

We thank all those who have lent their support, and especially Nicolas Mangold, Michel Capderou, Thérèse Encrenaz, Emmanuelle Gautier, Azar Khalatbari and Sandrine Maïsano.

We are grateful for the many new images incorporated into this edition: sources include ESA and the HRSC and OMEGA teams, NASA (JPL and GSFC), and the associated teams of Malin Space Science Systems, Arizona State University, Cornell University, and the Lunar and Planetary Laboratory.

We also thank D. Andersen, B. Harris and D. Hardy, for allowing the use of their illustrations, and special thanks go to Kees Veenenbos, who reworked some of his extraordinary images (see www.space4case.com) especially for us. We are grateful to Luce Delabesse for technical assistance.

We express our gratitude also to the team at Éditions Belin, and notably Roman Ikonicoff and Yolande Limousin, for their remarkable work on the realisation of the French edition of this book – not forgetting the contribution of Christelle Sauvage to its origins. During the preparation of the first edition, the authors were privileged to be able to work in the marvellous conditions provided by the Fondation de Treilles.

We are grateful to Clive Horwood and Praxis Publishing for their support and encouragement, and to Bob Mizon for the excellent translation work.

Lastly, we must thank our laboratory colleagues, researchers, engineers and technicians, and all those who have worked alongside us with the aim of bringing to fruition a European Mars exploration programme. It is to be hoped that their efforts will realise the dream of an international presence on Mars in the twenty-first century.

The publishers thank Roman Ikonicoff, who re-read and edited the text.

NASA's Hubble Space Telescope snapped this picture of Mars on 28 October 2005, within a day of its closest approach to Earth. A large regional dust storm, about 1500 km across, appears as the brighter, redder cloudy region in the middle of the planet's disk. The south polar ice cap is much smaller than normal because it has largely sublimated with the approach of southern summer. Bluish water-ice clouds can also be seen along the limbs and in the north (winter) polar region at the top of the image. (Courtesy NASA, ESA, The Hubble Heritage Team (STScI/AURA), J. Bell (Cornell University) and M. Wolff (Space Science Institute).)

Welcome to Mars!

After a journey of 300 million kilometres from Earth, the heat shield of the Mars Exploration Rover Opportunity crashed onto the surface of Mars on 25 January 2004. The shield had protected the precious payload as it entered the

atmosphere, before being detached. Opportunity parachuted down, landing a few hundred metres away. One year on, Opportunity, on its way southwards, encountered its protector, and transmitted this image. (Courtesy NASA, JPL (Jet Propulsion Laboratory), Cornell.)

Introduction

A simple red dot observed in the sky by our ancestors, or the planet across which our modern robots trundle in their explorations: however we see it, Mars never ceases to thrill.

Earth's sister planet, that other world which may harbour extraterrestrial beings, leading lives parallel to ours... Such a view of Mars was indeed the fruit of over-stimulated imaginations and a few sparse data, and today's probes, revealing the true nature of the planet, have dashed most of the ancient myths. However, recent findings have sprung their own surprises, and Mars is, if anything, an even more fascinating place than it was before.

What have we learned from the Mars probes? Mars, like planet Earth, is a world, a complex and vast world with a long history. This book will confirm these ideas, going as it does to the heart of current planetary research, with all its doubts (and occasional certainties). We present many images, some of them previously unpublished. Although Mars is a beautiful place, a book about it need not be a mere photo album. The images will be markers along the way as we reveal this new world which, although it has many echoes on our planet, is nonetheless quite 'extraterrestrial'.

We shall discover Olympus Mons*, more than 20,000 metres high and the solar system's biggest volcano. Beside Olympus, Everest would rank as just a foothill. At Mars' poles, glaciers, formed from thousands of fine strata, are evidence of past climatic fluctuations. Entire plains bear the marks of cataclysmic flooding. Everywhere, strange dunes whisper the story of the winds that shaped them...

Imagine a day spent somewhere on Mars. As dawn breaks, night frosts give way to mists and fogs which, in their turn, disperse to reveal the orange sky. At the warmest part of the day, dust devils patrol the surface. At dusk, a tiny, oddly shaped moon takes just a few hours to cross the sky. And tomorrow, will the whole planet be veiled beneath a global dust storm, as happens sometimes in the winter?

Beyond the magnificent scenery, there is a new world to understand, and the workings of the planet to reveal. How was Mars formed? Why has its evolution followed a different path to that of Earth? What do its river beds, volcanoes and glaciers tell us about its past? Could life have existed there? Does it exist there now? What processes 'drive' Mars today?

* Most of the places mentioned in the text are shown on the map of Mars at the end of the book.

Mars as seen from 55.8 million kilometres by the Hubble Space Telescope in orbit around the Earth on 26 August 2003. On that day, Mars was closer to the Earth than it had ever been during the whole of human history (in fact, since 60 000 years ago). It is the end of spring in the southern hemisphere of Mars. The last deposits of frozen carbon dioxide, accumulated during the winter around the south pole (bottom), are subliming in the spring sunshine. At the near edge, a tongue of ice dubbed the 'Mitchell Mountains' (though there is nothing mountainous about it) projects from the ice cap. In the southern hemisphere, spring is a relatively warm season, and few clouds are present in the atmosphere. Only the north polar region (at top) is shrouded in haze. (Courtesy NASA, J. Bell (Cornell U.) and M. Wolff (SSI).)

Of all the planets of the solar system, Mars is the one that most closely resembles ours. Geologists, meteorologists and many other scientists possess a vast 'laboratory' to test out their comprehension of the phenomena which occur on the terrestrial planets. Mars has the added advantage of having preserved the

A nest on Mars. The little crater 'Eagle' in the Terra Meridiani region. The six-wheeled rover Opportunity fortuitously landed in the middle of this crater on 25 January 2004. Sixty days later, after descending from its landing platform (left of centre) and investigating nearby rocks and soil, Opportunity leaves the nest and sets off for new adventures (see pages 201–204). (Courtesy NASA, JPL, Cornell.)

traces of its distant past. The same cannot be said of the Earth. Our planet's surface is constantly renewed: erosion and the phenomenon of plate tectonics have erased the archives of four-fifths of Earth's history. There remain almost no clues to what the Earth was like when the first micro-organisms made their appearance, before carbon chemistry was so widely exploited by living things. Mars can remind us of our own past, and perhaps tell us how life began. To understand Mars is, in a sense, to understand our own origins.

Let us leave Mars for a moment and return to Earth. All over the world, hundreds of researchers and engineers have but one aim in mind: bringing to fruition future Mars missions. In the last decade, more than ten space probes and their robot explorers have been sent to the red planet; the next twenty years will see more and more voyages from Earth to Mars. There will be new probes, and samples of Martian surface material will be brought back. Perhaps people too will make the journey.

The five parts of the present book trace the history of this active planet. Part 1 examines its formation, together with that of the solar system itself, from the ashes of dead stars, more than 4.5 billion years ago. Part 2 takes us through its early and turbulent youth. Part 3 traces the gradual, 3.5-billion-year long metamorphosis which created the great planetary structures. Part 4 explores Mars as we find it today, with its dust storms, water features and atmosphere, and we see that Mars is subject to continual climatic change. Finally, in Part 5, we recount the story of the recent exploration of Mars, and look at what is going on in laboratories and space agencies in preparation for the missions of the next twenty years.

This new edition offers a synthesis of our current knowledge about Mars, revisited with the aid of the multitude of images and discoveries made by Mars Express, the first European probe, and by NASA's robot geologists, the indefatigable Mars Exploration Rovers.

So, *bon voyage*, Mars traveller: don't forget that Mars still holds many secrets, laying down a challenge for generations to come...

Comparison of Mars and the Earth

	Mars	Earth
Size (mean diameter)	6779 km	12742 km
Age	Formed 4.5 bn yrs ago 50% of surface more than 3.8 bn years old	Formed 4.5 bn yrs ago 99% of surface less than 2 bn years old
Position	Eccentric orbit Distance from Sun: 1.384–1.664 AU*	Near-circular orbit Distance from Sun: 0.983–1.017 AU
Data	**Mass**: $6.42.10^{23}$ kg (0.107 Earth mass) **Surface gravity**: 3.72 $m.s^{-2}$ (0.38 Earth's) **Day length**: 24h 39m 35s **Year**: 669 Mars days (=687 Earth days)	**Mass**: $5.97.10^{24}$ kg **Surface gravity**: 9.81 $m.s^{-2}$ **Day length**: 24h **Year**: 365.25 Earth days
Atmosphere	**Colour**: orange **Composition**: CO_2 95.3%, N 2.7%, Ar 1.6%, O 0.13% **Mean surface pressure**: $\sim$6 hPa **Other**: assorted mineral dust in suspension Some water-ice & CO_2 clouds	**Colour**: blue **Composition**: N 78.1% O 20.9%, Ar 0.93%, Water vapour $\sim$1% **Mean surface pressure**: 1000 hPa **Other**: many water clouds Protective ozone layer
'Family'	2 small satellites: Phobos (size: 27 × 22 × 18 km, 5990 km above Mars' surface) Deimos (size: 15 × 12 × 10 km, 20,100 km above Mars' surface)	One large satellite: the Moon (diameter: 3476 km, approx. 378 000 km from Earth)

*1 AU (Astronomical Unit) = 149.6 million km
(Data taken from NSSDC, NASA Goddard Space Flight Center)

Martian landscapes

The plain within Gusev Crater, seen from the Columbia Hills by the Mars Exploration Rover Spirit in August 2005. (Courtesy NASA, JPL-Caltech, Cornell.)

Outcrops of rock rich in sulphate salts, in the Terra Meridiani region, photographed by the Mars Exploration Rover Opportunity in September 2005. (Courtesy NASA, JPL-Caltch, Cornell.)

An expanse of sand traversed by Opportunity in May 2005. (Courtesy NASA, JPL, Cornell.)

The flank of Endurance Crater, surveyed by Opportunity, June-December 2004. (Courtesy NASA, JPL, Cornell.)

A cliff of layered rocks on a promontory called Cape St. Mary, examined by the Opportunity rover as part of its investigation of the rim of Victoria crater. This view combines several images into a false-colour mosaic that enhances subtle colour differences in the rocks and soil. (Courtesy NASA, JPL-Caltech, Cornell.)

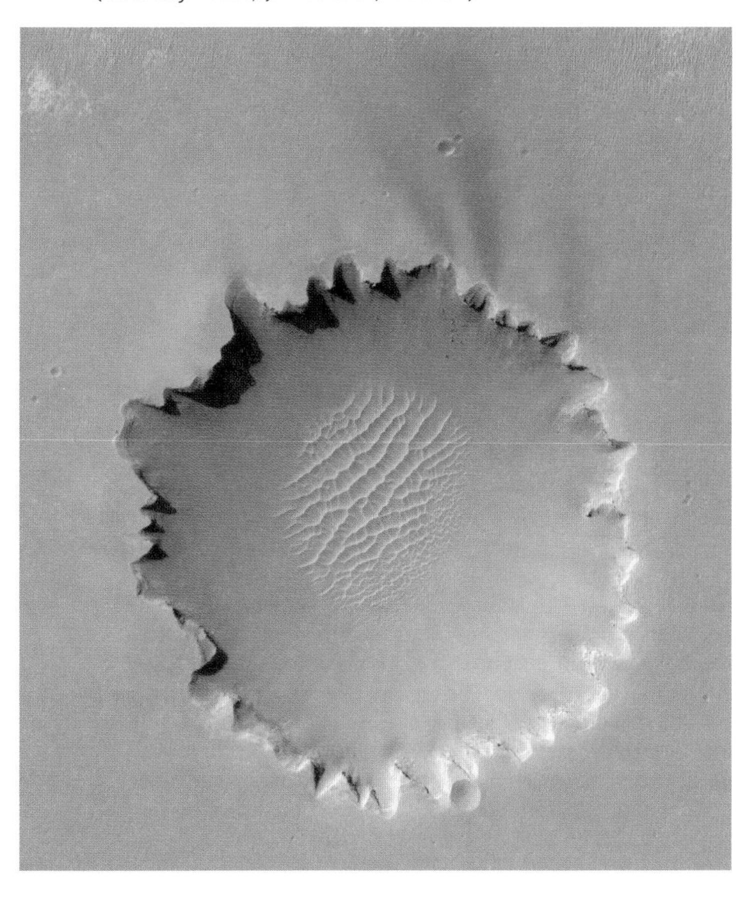

Left: Victoria crater, an impact crater on Meridiani Planum, viewed from the HiRISE camera on Mars Reconnaissance Orbiter. The crater is approximately 800 metres in diameter. It has a distinctive scalloped shape to its rim, caused by erosion and downhill movement of crater wall material. (Courtesy NASA, JPL-Caltech, University of Arizona.)

Chronology of Mars

To date Martian terrains, geologists count meteorite impacts ... and visit the Moon

Before undertaking any in-depth study of the geophysical evolution of a planet, scientists must possess a time-scale within which to classify the various structures observed. How, though, can we date the Martian terrains if we have no samples of the surface? The answer lies in counting craters, formed over long periods by incoming meteorites. All dates for geological structures used in the present book are based on this principle.

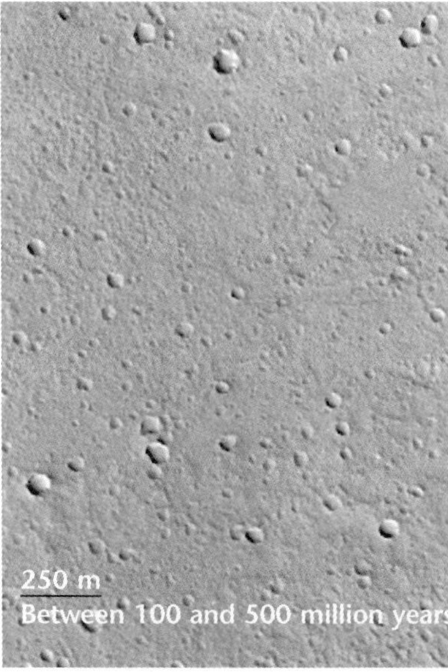

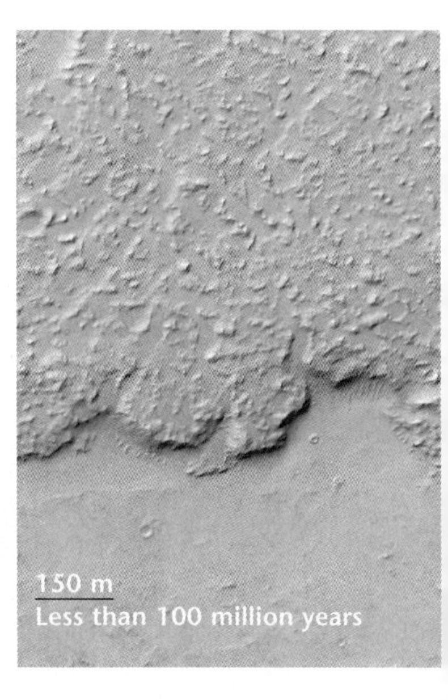

Unequal cratering. Left: a surface peppered with impact craters in Terra Arabia, exposed for the last three billion years. Many craters hundreds of metres in diameter are seen. The smallest craters have been eroded or covered in sediments. **Centre**: the surface of the caldera (volcanic crater) of the volcano Arsia Mons, in the Tharsis region. The innumerable small craters suggest an age between 100 million and 500 million years. **Right**: high-resolution image of a recent lava flow (at top) in the region of Daedalia Planum, to the south-west of the volcano Arsia Mons. Note the absence of impact craters, suggesting that this is a relatively young surface, less than 100 million years old. Illumination is from the left on all three images. (Courtesy NASA, JPL, Malin Space Science Systems.)

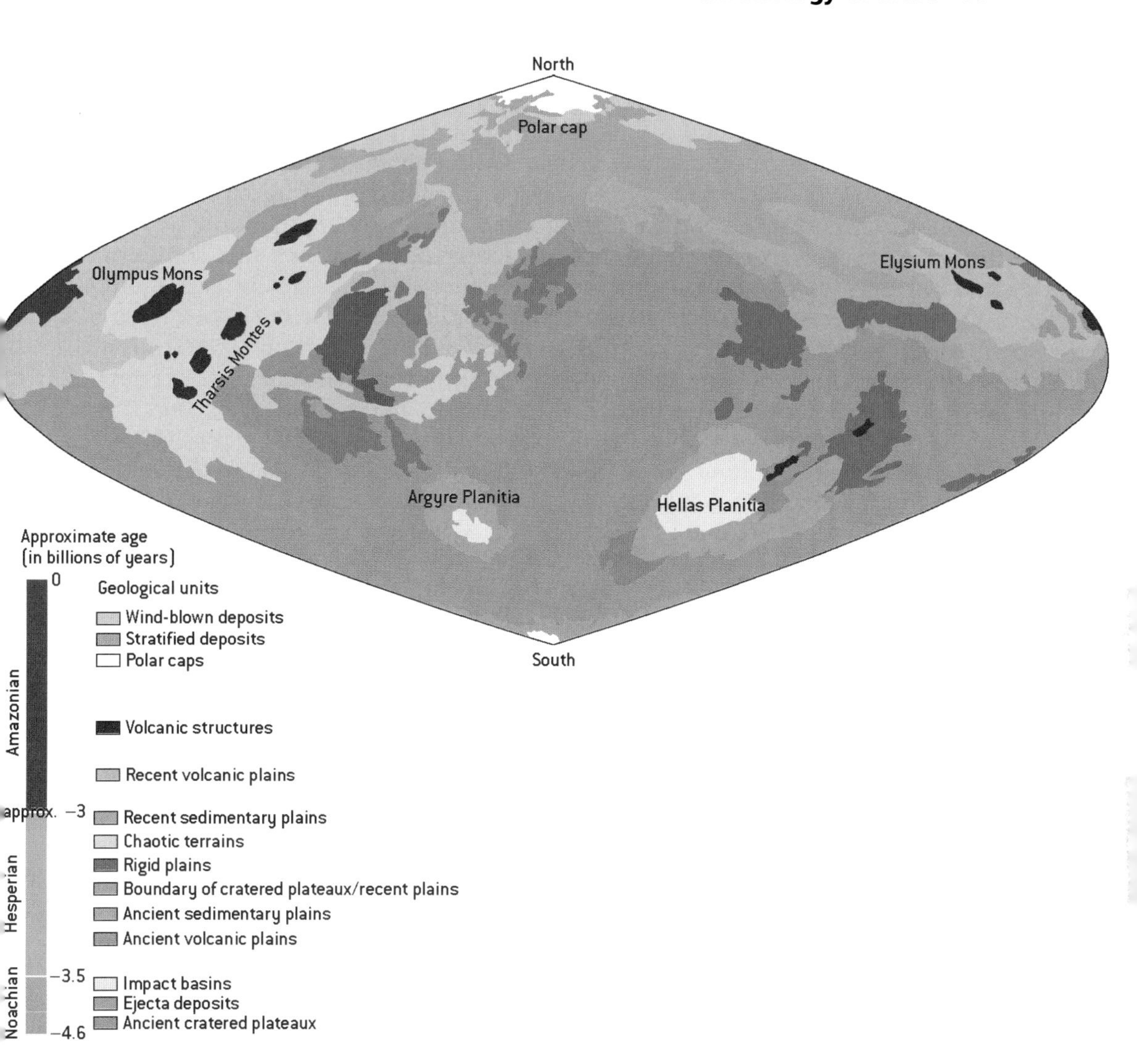

Map of the principal geological areas on Mars. Dating by cratering has led to the definition of the three main geological eras of the planet. The most ancient surfaces on Mars, such as the cratered plateau of its southern hemisphere, date from the Noachian era (–4.5 to about –3.5 billion years). The Hesperian era saw the formation of the northern plains and certain areas of the cratered plateau. The most recent era was the Amazonian, represented by the plains where the last phases of volcanic activity occurred, and by polar caps and channels. The date of transition between the Hesperian and the Amazonian remains controversial, but lies somewhere between 2 and 3.2 billion years ago.

Not long after it was formed, 4.5 billion years ago (see Part 1), Mars was subjected to a continuous meteoritic bombardment. Intense until 3.8 billion years ago, this process rapidly diminished thereafter (the impactors decreasing in both number and size), and stabilised over the last billion years. This decrease in the frequency of impacts over time allows geologists to establish a relative chronology from calculations involving the density of craters per unit area on the surface. For example, the heavily cratered plateau of the southern hemisphere, which will be referred to quite often and especially in Part 2, has large numbers of craters of considerable diameter. It must therefore be older than any lightly cratered surface such as is found in the plains of the northern hemisphere. However, counting craters gives only the relative age of a region compared with another. So although we might say that the high plateau in the south is 'definitely' older than the northern plains, to know its exact age, we would need a sample from the Martian surface in order to be able to date it by measuring radioactivity (see page 31). Although it is not currently possible to return the precious samples, future space missions will be programmed to do this, as will be shown in Part 5. Meanwhile, planetologists have to resort to a kind of 'short-cut', involving lunar samples brought back by the Apollo astronauts. Our own natural satellite also possesses, of course, a 'cratering curve'. The lunar samples serve to calibrate this curve, substituting absolute dating for relative dating. Mars specialists have been able to do the same thing, calibrating their calculations for the ages of areas on Mars using lunar samples, and adding 'correction factors' which take into account the position of Mars relative to the Sun, Martian gravity, the presence of an atmosphere, etc. Until we possess a better method, this extrapolation is sufficient for our purposes, even if absolute estimates differ a little from each other according to the values allotted to the corrections, and the models used.

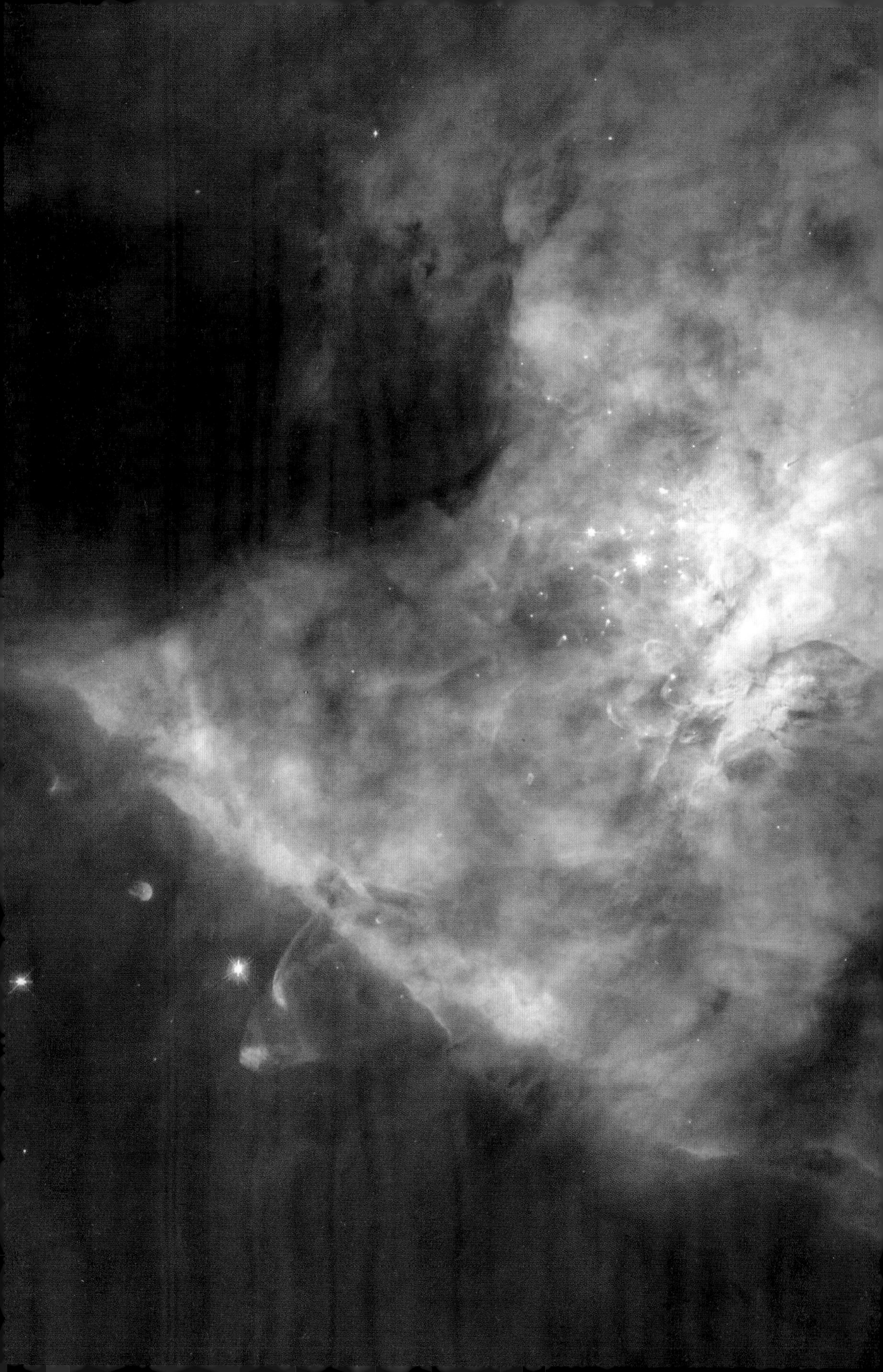

Birth of a planet

–4.6 to –3.8 billion years

Our history of Mars begins in the darkness of past time, at an epoch when almost nothing of the solar system that we know existed – not even the Sun. But the universe was old even then, some 10 billion years old, two-thirds of its present age. All the matter that would form the Sun, Mars, the Earth and living things was still floating as gas and dust.

Suddenly, in just a few tens of millions of years, the blink of an eye on the cosmic time-scale, the solar system came into being. What followed would forever leave its mark on the destiny of the red planet.

The Orion Nebula, photographed by the Hubble Space Telescope. This immense cloud of matter may resemble the nebula which gave birth to our Sun, and to Mars and the other planets of the solar system. The Orion Nebula is about 1500 light-years away from the Sun, in the same spiral arm of our galaxy, the Milky Way. In the heart of the nebula, more than 150 recently-formed 'protoplanetary' discs, a few million years old, have been identified. Some of these will possibly give rise to Mars-like planets. The image is about 2.5 light-years across. (Courtesy NASA, STScI, C.R. O'Dell and S.K. Wong, Rice U.)

1 Before Mars and the other planets

What is the origin of the matter which formed the planets of the solar system?

Like the other 'terrestrial' planets, Mercury, Venus and the Earth, Mars is made principally of the so-called 'heavy' chemical elements, such as oxygen, magnesium, silicon and iron.

These elements were not created by the Big Bang, the singularity which marked the birth of the universe some 13.7 billion years ago. This primordial

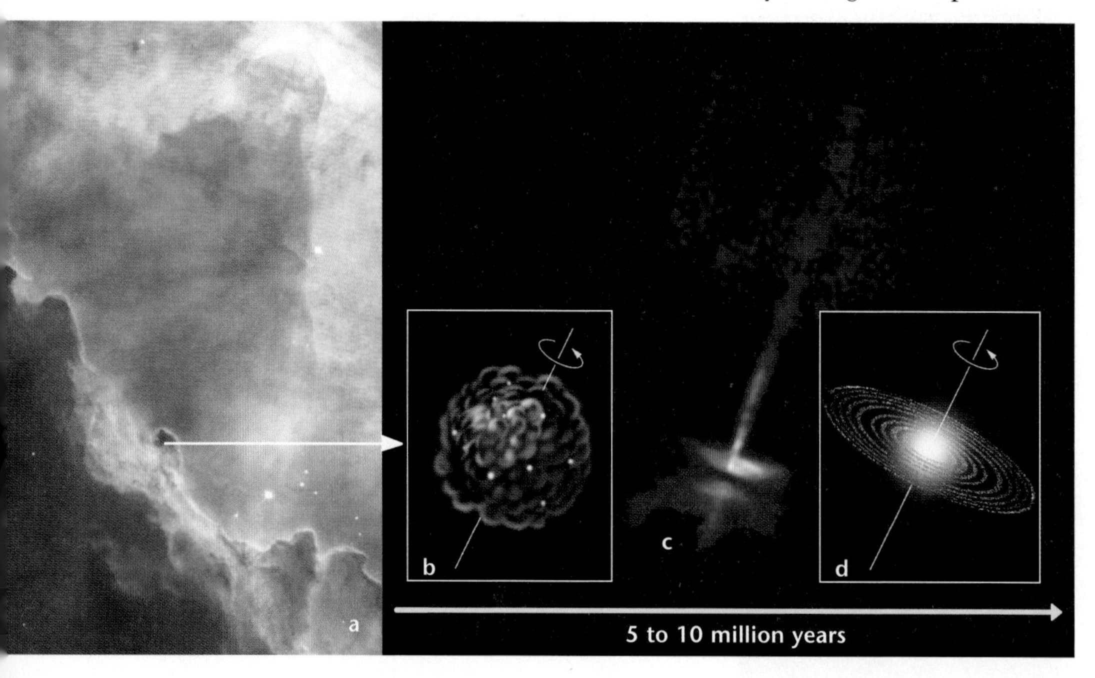

Stellar nurseries. Image (a) shows part of the 'Omega' or 'Swan' Nebula (M17). (Courtesy NASA, ESA and J. Hester (ASU).) This cloud of dust and gas is a veritable 'nursery' of stars. When temperature and pressure locally reach a critical threshold, embryonic stars appear. Then, starbirth proceeds, in three stages. Diagram (b): still surrounded by a residual disc of gas and dust, the young star continues to draw in matter though its own gravity, and then (c) begins to eject particles in a direction perpendicular to the plane of the gaseous disc, within an intense magnetic field. This phenomenon slows down the rotation of the star. Here we see the example of HH30, less than half a million years old: its disc of matter appears dark, and its perpendicular jet bright. (Courtesy C. Burrows (STScI & ESA), the WFPC 2 Investigation Definition Team, and NASA.) In (d), the disc, containing only a fraction of its initial mass, now possesses most of the rotational energy (angular momentum). Now, planetary formation may begin...

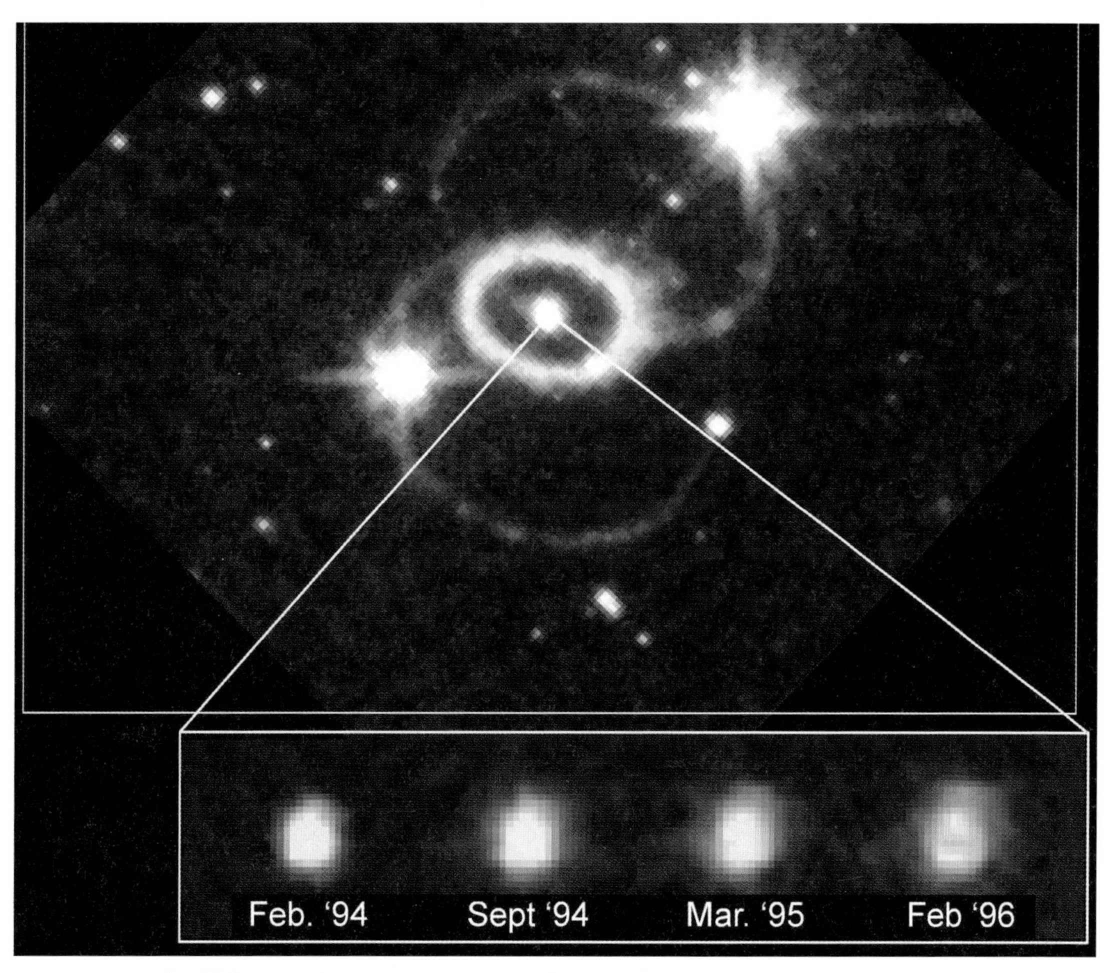

A cosmic tidal wave. Supernova SN1987A, photographed by the Hubble Space Telescope. The supernova occurred in the Large Magellanic Cloud, a satellite galaxy of our own Milky Way. In 1997, ten years after it exploded, SN1987A's expanding wave of material had travelled more than 750 billion kilometres (5000 times the distance of the Earth from the Sun), at an average velocity of 2500 km/s. Scientists believe that a wave of matter such as this triggered the formation of the solar system 4.56 billion years ago. This wave is rich in very unstable radioactive elements, such as aluminium-26. Now, within the most ancient meteorites in the solar system, magnesium-26, a by-product of the disintegration of aluminium-26, has been discovered. In this way, a link has been established between the formation of the solar system and the explosion of a supernova. (Courtesy Chun Shing, Jason Pun (NASA/GSFC), Robert P. Kirshner (Harvard-Smithsonian Center for Astrophysics), and NASA.)

explosion produced only the very lightest elements, such as hydrogen, helium and a little lithium and deuterium. So where did the planet-forming elements originate?

One very early hypothesis claimed that the heavy elements in planets were made by the Sun. This idea is thought to have originated in the mind of the

Greek philosopher Anaxagoras (500–428 BC), who said that the existence of iron meteorites proved that the Sun was nothing but 'a mass of red-hot iron'.

Nowadays we know that the Sun is composed essentially of hydrogen and helium, but the idea that the Sun provided the material for the formation of the planets is not such an absurd one, because those heavy elements are indeed present within it. What is more, they are present in proportions quite close to those in the primitive meteorites that built the planets. For example, in the Sun (just as in these meteorites), carbon atoms are eleven times more numerous than silicon atoms. However, the hypothesis of a solar origin for the planets encounters one major objection: it cannot account for the disparity of the amount of light elements, especially deuterium, lithium and boron. For when the planets formed, the Sun was deficient in these elements, since it was burning them (and it still is) – and planetary material does not exhibit such deficiency.

Today we can give a better answer to the question of the origin of planets, thanks to our analysis of ancient solar system meteorites. The Sun and its planets owe their heavy elements to the explosions of massive stars at the end of their lives. As a star enters the final stage of its life, having transformed all its hydrogen into helium, it produces heavy elements: helium is transformed into carbon, oxygen and nitrogen, and these in their turn become still heavier elements as fusion processes continue. In stars with sufficient mass, this chain of creation ends in a supernova explosion, involving the ejection of all these elements into space. They will then participate in the formation of new stars and planets.

2 Very rapid protoplanetary growth

From stardust to planet in 100 million years

4.52 billion years ago, the Sun began to shine within a disc of gas and dust. Near it, the temperature was so high that water and other volatile elements were in a gaseous state. Only compounds which were not very volatile, such as oxides of silicon and magnesium, and metallic elements, were able to remain in solid form. Dust particles collided and adhered to form porous agglomerates, about a metre in diameter. In their turn, these bodies collided with each other, until they became kilometre-sized 'planetesimals', the building blocks of future planets.

After a fairly tranquil 'infancy', the Sun entered a violent adolescent phase at about the age of 10 million years. For the first time since their formation, gases and volatile materials reversed their movement, now being ejected towards the edge of the solar system due to the buffeting of the solar wind, an intense stream of particles issuing from the Sun's corona.

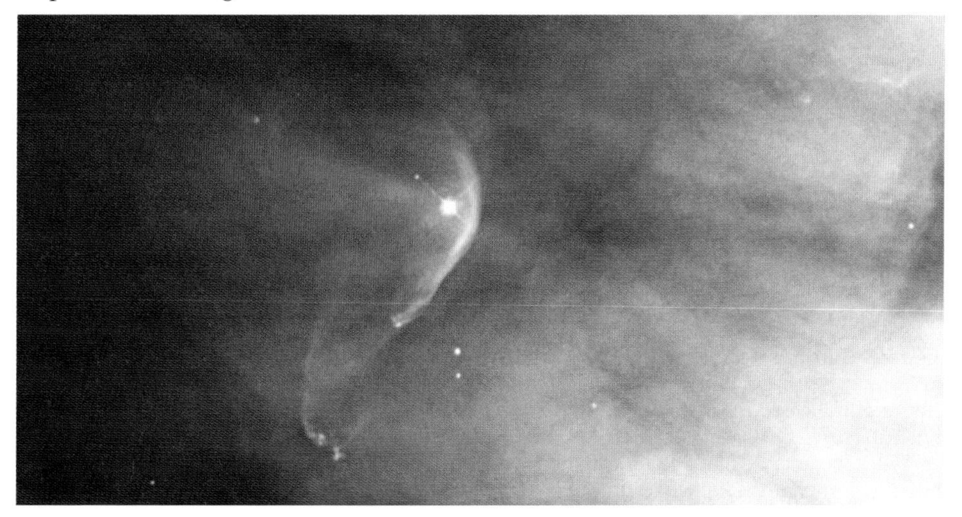

In the Orion Nebula, the young star LL Ori (at the centre of the image) suggests what the Sun probably looked like during the 'T Tauri' phase of its adolescence. This phase is characterised by the existence of a fierce solar wind, and the ejection of a quantity of matter of the order of one percent of a solar mass every million years. These phenomena can be from 100 to 1000 times more intense in the so-called 'FU Ori' phase. The bright zone of the nebula (to the right of LL Ori in this image) corresponds to a region where hydrogen, oxygen and nitrogen atoms are being excited most strongly by the solar wind. (Courtesy NASA and The Hubble Heritage Team (STScI/AURA).)

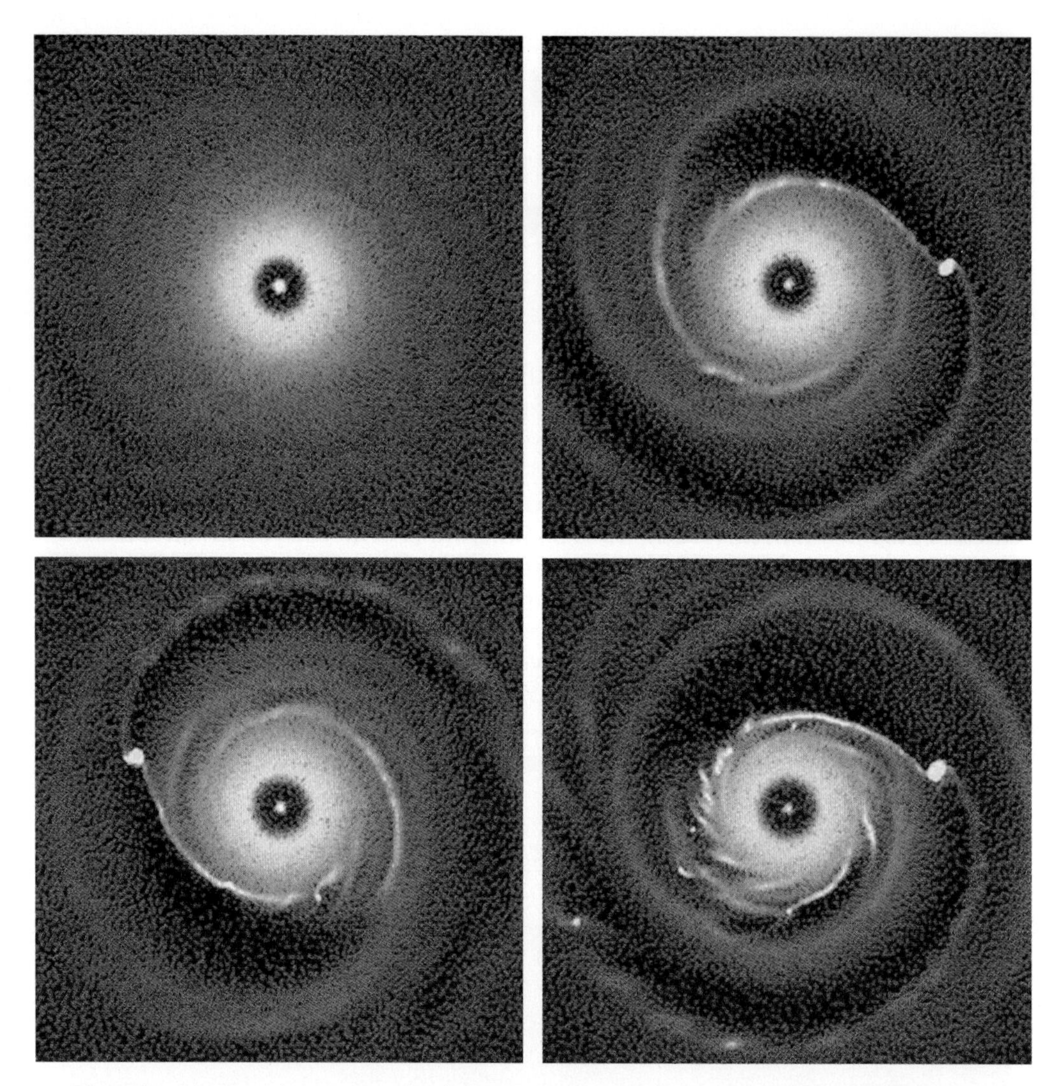

The role of Jupiter. A computer simulation by P. Armitage, of the University of Toronto, illustrates the process of planetary formation. 60 000 small bodies were initially 'placed into orbit' around a 'Sun'. As shown in figure (a), without the presence of a massive planet, the configuration evolves very slowly (the simulation represents a period of ten million years). If a Jupiter-type planet of mass one-thousandth that of the Sun is introduced into the system, the configuration after the same length of time corresponds to figure (b): the small bodies are concentrated as much between 'Jupiter' and the 'Sun' as outside the orbit of 'Jupiter'. The other figures represent the process of aggregation after phase (b), over two million (c) and four million (d) years. The embryos of future planets are evident. If the simulation were pursued, we would notice that 'Jupiter' prevents any further formation of large planets within the asteroid belt, and limits the growth of a 'Mars' to its present size. In summary, in the absence of Jupiter, Mars might have become a little bigger, but the Earth and Venus would probably not have been able to attain their current dimensions. What might then have become of these planets? (Courtesy P.J. Armitage and B.M.S. Hansen, *Nature* No 402, 9 Dec. 1999, p. 20.)

As matter flowed outwards, water vapour migrated to regions where temperatures were lower. At about 820 million kilometres from the Sun (5.5 times the present distance between the Earth and the Sun), water vapour condensed into particles of ice. The ice came together with dust to form a planet several tens of times more massive than the current mass of the Earth. The gravitational field around this massive body drew in large quantities of hydrogen: Jupiter was born, and it had taken less than 10 million years.

Saturn took twice as long to form, at a distance from the Sun (about 1250 million kilometres) where its orbital period was twice that of Jupiter. Uranus and Neptune were born in similar circumstances, devouring icy planetesimals at a distance from the Sun more than twice that of Saturn.

Nearer to the Sun, planetesimals continued to attract each other and collide. In the course of about 10 million years, embryonic planets grew into the smaller planets of the early solar system. Some were the same size as Mars, about 7000 kilometres in diameter, about 10 of them resembled Mercury (5000 kilometres), and there were many hundreds similar in diameter to our Moon (3500 kilometres). So, between 4.47 and 4.44 billion years ago, the 'modern' phase of the history of the solar system began. As the last cataclysmic collisions between planets occurred, the Earth and Venus gathered in most of the existing material. Another planet, as big as Mars, had formed, and collided with the Earth. From this impact was born the Moon. Later, Jupiter moved closer to the Sun, causing Saturn and, further out, Uranus and Neptune, to move away. Since then, the planets have maintained the orbits which they pursue today.

3 Formation of the core, crust and mantle

The physics of the blast furnace at the heart of planets

The 'terrestrial' planets (Mercury, Venus, the Earth and Mars) were formed from a disorderly crowd of planetesimals. However, these planets now exhibit an iron core surrounded by a rocky mantle, which is not so dense, and above this lies an external crust. This differentiation could not have happened unless the interior of the primitive planets had melted, and followed the same process of refinement as occurs in high-temperature industrial furnaces. During the 1970s, most planetologists thought that, in the case of Mars, this process took a billion years, which is the time required to accumulate a sufficient quantity of heat created by the disintegration of radioactive elements within the planet. However, analysis of Martian meteorites during the 1980s (see pages 30–32) revealed a much more rapid scenario. In these meteorites, which bear witness to the formation of Mars' crust, certain elements known for their readiness to associate with iron ('siderophilic' or iron-loving elements), and which are found with iron even within the core, were not present in the expected quantities. This gave credence to the idea that iron had largely migrated from the crust of Mars towards the core as the planet accreted, the core having formed in fewer than 50 million years, and possibly only 20 million years.

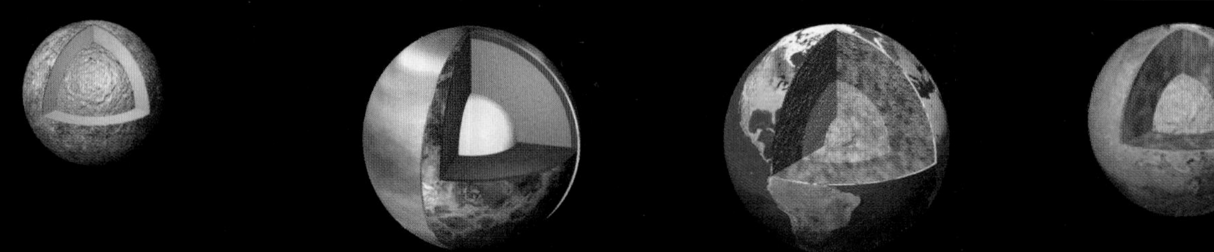

Mercury, Venus, the Earth and Mars: the four terrestrial planets all possess a metallic core, a mantle and an external crust. They are represented here with details of their radii (R), masses (m) and mean distances (d) from the Sun. The comparatively enormous core of Mercury is exceptional. This planet was probably involved in a vast collision towards the end of its formation, in the course of which the core of the impactor penetrated the mantle of the proto-planet and fused with its core. Most of the fragments of the mantle were ejected. (Courtesy Calvin J. Hamilton.)

During the accretion phase of a planet, the impacts of planetesimals liberate vast quantities of energy. An object 10 metres in diameter, crashing into the Earth at 20 kilometres per second (a typical impact velocity for a meteorite), liberates energy equivalent to 100 000 tonnes of TNT, or about 7 times the energy of the Hiroshima bomb. What would be the effect of a 1-kilometre planetesimal? The greatest impacts may convert up to 30% of their kinetic energy (energy of motion) into heat. This heat would be sufficient to melt the surface of a planet to a depth of several hundred kilometres. (Illustration by David A. Hardy.)

In fact, it was accretion itself that produced sufficient heat, not only on Mars but also on the Earth, Venus and other terrestrial bodies, to refine the elements from the original mixture present within the planetesimals, in the manner of a blast furnace. In the heart of a blast furnace, oxygen, carbon and iron oxides react to form iron and carbon dioxide: within the magma of a planet, iron oxides and graphite from meteorites known as carbonaceous chondrites react to form metallic iron and carbon dioxide. The molten iron sinks towards the centre of the planet, gaining a little more heat as it falls. Conversely, 'refractory' oxides, and in particular aluminium oxide and calcium oxide, rise to the surface as a kind of scum: the 'primary crust'.

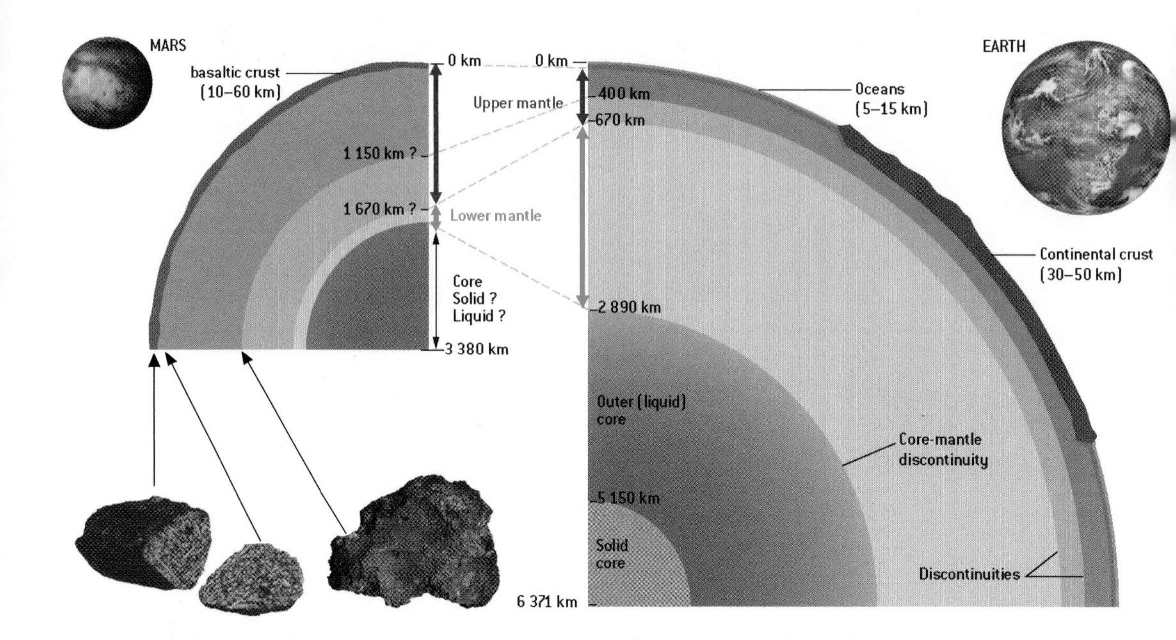

How big is the core of Mars? And what is it made of? The depth of Mars' core-mantle boundary has not yet been directly measured. However, we can have some indication by comparing meteoritic fragments of the Martian crust (such as the Chassigny meteorite) with meteorites with a composition supposedly comparable to that of planetesimals, such as the carbonaceous chondrite which fell near Ivuna, in Tanzania, in 1938. The core of Mars, which is about 1500–1900 km in diameter (about half the diameter of the planet), is probably composed of about 70-80 percent iron, with some nickel and sulphur. Having studied the SNC meteorites (see pages 30–32), Heinrich Wänke, of the University of Mainz in Germany, concluded that the sulphur content was 14 percent and the nickel content 8 percent. One of the consequences of the presence of sulphur in Mars' core is to decrease by nearly 300°C the solidification temperature of the iron-nickel-sulphur amalgam. So, because of the presence of sulphur, Mars could still, even today, possess a completely liquid core. Recent geodesic measurements by Mars Global Surveyor tend to confirm this. (Illustrations courtesy Sté Labenne Meteorites, France (left) and INSU, CNES (right).)

4 The evolution of Mars, the Earth, Venus and Mercury

Born at the same time – but their paths are so divergent

Various factors can be invoked to explain the evolutionary differences of the four terrestrial planets. Among these factors are initial size, the history of (cataclysmic) cratering, and the process of heat loss from within.

Initial size

By the end of the period of accretion, a little more than 4.45 billion years ago, Mars was half the size of Venus or the Earth, and about ten times less massive. This difference would prove decisive, in more ways than one, for the quantity of energy stored within the planets. The more massive a planet is, the greater is the

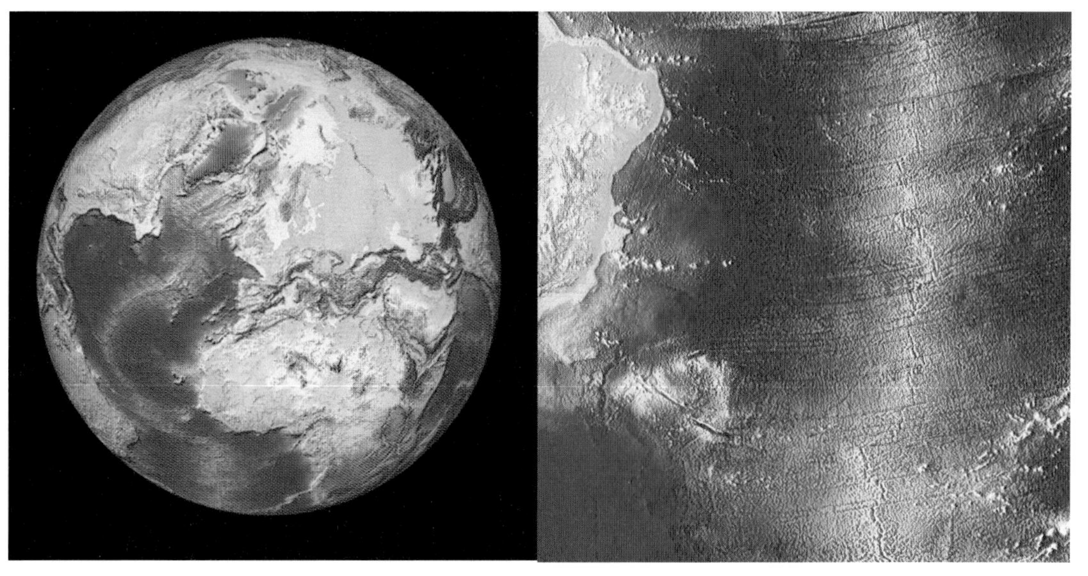

On Earth ... The Earth is characterised by plate tectonics, its continental and oceanic ('lithospheric') plates moving at a rate of a few centimetres per year. This movement at the surface is associated with convection currents within the mantle. This phenomenon accounts for two-thirds of the heat loss from the mantle and core. It is in the volcanic zones of oceanic ridges that the plates move apart, allowing hot matter from the depths to emerge and form new crustal material at the rate of 3 square kilometres of oceanic crust per year. On the left is shown the Atlantic Ridge, and on the right, an enlargement of its southern part. This ocean ridge volcanism extends for more than 60 000 kilometres. In subduction zones, the ancient crust falls back into the mantle, to be recycled. (Courtesy NOAA, Satellites and Information.)

gravitational acceleration towards it of planetesimals, and the harder they impact as accretion proceeds. Also, the bigger the planet, the greater the quantity of radioactive elements stored within it. What is more, large planets conserve heat better, since their stronger gravity can maintain dense atmospheres, opaque to thermal radiation and therefore limiting energy loss. This is why, from the beginning, the two smaller bodies, Mercury and the Moon, cooled more quickly. Mars marks an intermediate stage between these two smaller bodies and two larger ones, the Earth and Venus.

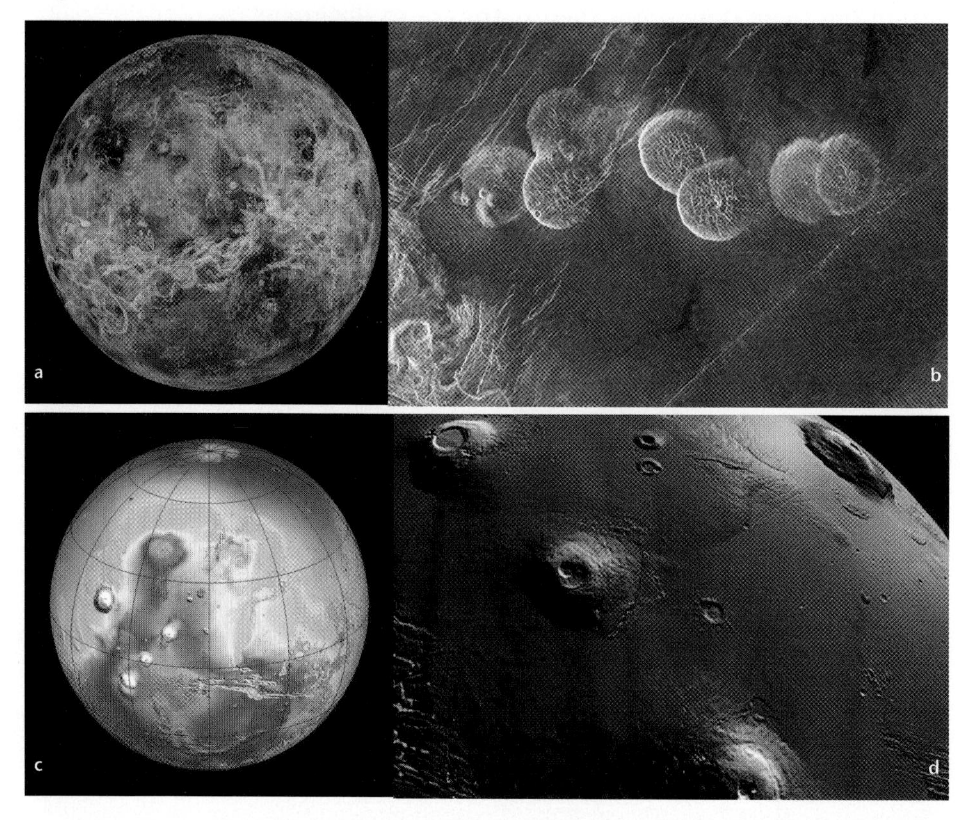

... And on Venus and Mars. Unlike the Earth, Venus and Mars show no plate-tectonic activity. Their internal heat is probably lost through local up-currents in the mantle (volcanoes), forming cylindrical plumes at the surface. On Venus (image (a)), seven pancake dome volcanoes have attained heights of 750 metres (b) in the Alpha Regio area (Magellan radar image). On this planet, heat accumulated beneath a layer of rigid rock, the lithosphere. In consequence, large numbers of volcanoes erupted between 500 and 300 million years ago, covering most of Venus' surface with lava, and erasing the majority of surface impact features. As for Mars (c), it could well still be volcanic, especially in the Tharsis Region. Image (d) is a false-colour photograph of the topography of the Tharsis region, the vertical scale having been exaggerated by a factor of three. In fact, since the red planet has a very thick lithosphere (of the order of five hundred kilometres), thermally isolating the mantle and the core, it loses its heat much more slowly than does the Earth. Also, unlike the Earth's core, which is already partly solid, Mars' core is probably still in the liquid state. (Courtesy NASA (a) and (b), NASA, MOLA Science Team (c), and NASA, Goddard Space Flight Center, Scientific Visualization Studio (d).)

Cataclysmic cratering

The random process of cataclysmic meteorite impacts played an important part in the internal evolution of planets. Take the example of the Earth, whose mantle absorbed the greater part of the matter and energy involved in the violent impact which gave birth to the Moon, just over 4.48 billion years ago. Compare this with Mercury, which lost most of its mantle to another impact.

The process of heat loss

After the phase during which they absorbed material and gravitational energy, the planets radiated away some of their internal heat, including the heat slowly released by the radioactivity of the rocks. What mechanisms were involved? In the case of a solid, rigid planet, the process will be that of thermal conduction of heat towards the surface. This mechanism causes heat to be lost in proportion to the surface area of the planet, while heat retained is proportional to the volume. So, a large planet will lose relatively less energy than a smaller one. However, this model does not take into account the thickness of the isolating surface layers, volcanism or the movements of material within the mantle. What happens within the seemingly solid mantle is that, over geologic time, the hot rocks are subject to deformation, and material from deep within, hot and less dense, can come to the surface, where it cools. This phenomenon, known as convection, is much more efficient than conduction at transporting and releasing heat. The Earth is today the terrestrial planet with the strongest convection, almost up to the surface at the mid-oceanic ridges, while mantle convection has probably ceased in the Moon and Mercury. Again, Mars is in-between, with a mantle hot enough to convect weakly or at least, especially beneath the large volcanoes, with zones of partial melting, generate volcanism in the last hundred million years.

5 What Martian meteorites tell us

Stones reveal the secrets of the planet and its history

In the morning of 3 October 1815, the inhabitants of the village of Chassigny, in eastern France, were startled by some very unusual detonations: a 'stone' had fallen from the sky nearby. This 4-kilogramme meteorite intrigued generations of researchers, seeming to have come straight from the mouth of a volcano!

Of all the tens of thousands of meteorites found on Earth, only about thirty are of the 'Chassigny' type. These are known as the SNC meteorites, referring to the initial letters of the places where the first three were found: Shergotty in India, Nakhla in Egypt and Chassigny in France. Only in the 1980s was it finally discovered that these meteorites originate on Mars. They are characterised by their 'isotopic' oxygen content. Oxygen has various isotopes: its nucleus has

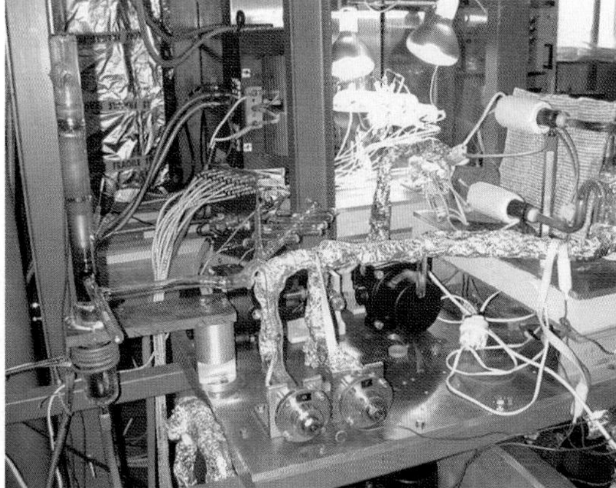

Discovered in the Antarctic, dated in the laboratory. Many Martian meteorites (e.g. EETA 79001) have been found on the ice of Antarctica (left), where they are easily seen. (Courtesy Antarctic Search for Meteorites Program (William Cassidy).) Their composition can be readily discovered by means of mass spectrometry (the right-hand photo shows equipment at the Institut de Physique du Globe in Paris).

To determine their origin, and more particularly the date at which they solidified from magma, geochemists rely on the phenomenon of the radioactive disintegration of potassium-40 into argon-40. As soon as the rock is molten, the argon present due to the disintegration of potassium-40 can escape. As soon as it has solidified, the argon accumulates within the rock. By comparing the content of different isotopes of potassium and argon with their initial values (their concentration in the Martian atmosphere), the time of crystallisation can be deduced. In the case of EETA 79001, the value is 180 million years. It is therefore quite a recent rock, given the age of the planet.

Clues to the differentiation and solidification of a planet. Isotopic geochemistry provides global data about a planet from only a few grammes of rock, using the properties of the many chemical elements in that rock. The basis of this analysis relies on two physical phenomena.
1: chemical elements (whatever their identity) usually possess several isotopes, i.e. atoms whose nuclei have the same number of protons but different numbers of neutrons. Certain isotopes are unstable, because of the number of neutrons within them. These will disintegrate more or less rapidly into other, more stable chemical elements.
2: certain chemical elements are fixed in rocks at the time of their formation. Others are produced in the interior of the rocks as a result of radioactive disintegration of elements present since the origin of the rocks.
Example: the dating of the Martian crust. In a meteorite which has come from Mars' crust, we find two isotopes of neodymium: Nd-142 and Nd-144. We know that the Nd-142 present in the meteorite was produced by the disintegration of samarium-146 (Sm-146) which had been fixed in the crust since its formation, and of which only traces now remain. Its half-life (the time required for half the atoms of Sm-146 to disintegrate into Nd-142) is 'only' 100 million years. However, Nd-144 is not a product of any disintegration, and we know in what proportion the Nd-144 and Sm-146 occur at the time of their formation in the supernova which gave birth to the solar system (see pages 18–20).
Also, the proportion of Nd-142 compared with that of Nd-144 today allows us to deduce the proportion of Sn-146 relative to that of Nd-144 at the time of the formation of the crust. By comparing the latter to the initial proportions (in the supernova), we can work out the time which has elapsed since the birth of the solar system and the formation of the Martian crust. According to this analysis, the crust was probably formed 100 million years after the birth of the solar system. (Courtesy LPI, Houston and Calvin J. Hamilton.)

eight protons, but the number of neutrons can differ. The isotopic composition and the relative abundance of different isotopes are the same in all the SNC meteorites, providing a distinctive signature. Another identifying feature of SNCs is that they are relatively young, mainly magmatic or volcanic rocks. Isotopic dating reveals that the youngest of them crystallised less than 200 million years ago.

How was it discovered that SNC meteorites come from Mars?

Dating and analysis revealed only part of the scenario: the fact that they once belonged to a single, far-off parent body big enough to have produced magma and volcanism. Was it Mars? Venus? Or an ancient planet destroyed during the formation of the solar system? The SNCs had been ejected into space probably as the result of a violent asteroid impact, but relatively recently, so only existing planets can be considered. Finally, the most convincing evidence of a Martian origin came from one of these meteorites: EETA 79001. This object was collected in Antarctica in 1980. Inside it were bubbles of gas, trapped within dark flecks of basaltic glass. The glass had been formed at the time of the enormous impact which detached the rock from its planet. It was concluded that these gaseous inclusions were almost identical in composition to Mars' atmosphere, as analysed *in situ* by the Viking landers in 1976. On this basis, SNC meteorites are considered by many researchers to be true fragments of the red planet, and clues to its formation.

6 The origin of atmospheres and water

A composite and elusive origin

Mars, the Earth and Venus all have atmospheres essentially composed of nitrogen, oxygen, carbon dioxide, water vapour and rare gases such as helium, argon, neon, krypton and xenon. These atmospheres are distinguished from each other by the relative proportions of the various gases, and by their global pressures. For example, the pressure at the surface of Mars is between 100 and 150 times lower than on Earth, which in its turn has an atmospheric pressure 90 times lower than that on Venus.

Mars (left) and Venus (right) as seen through the Earth's atmosphere from the space shuttle Discovery. (Courtesy NASA.)

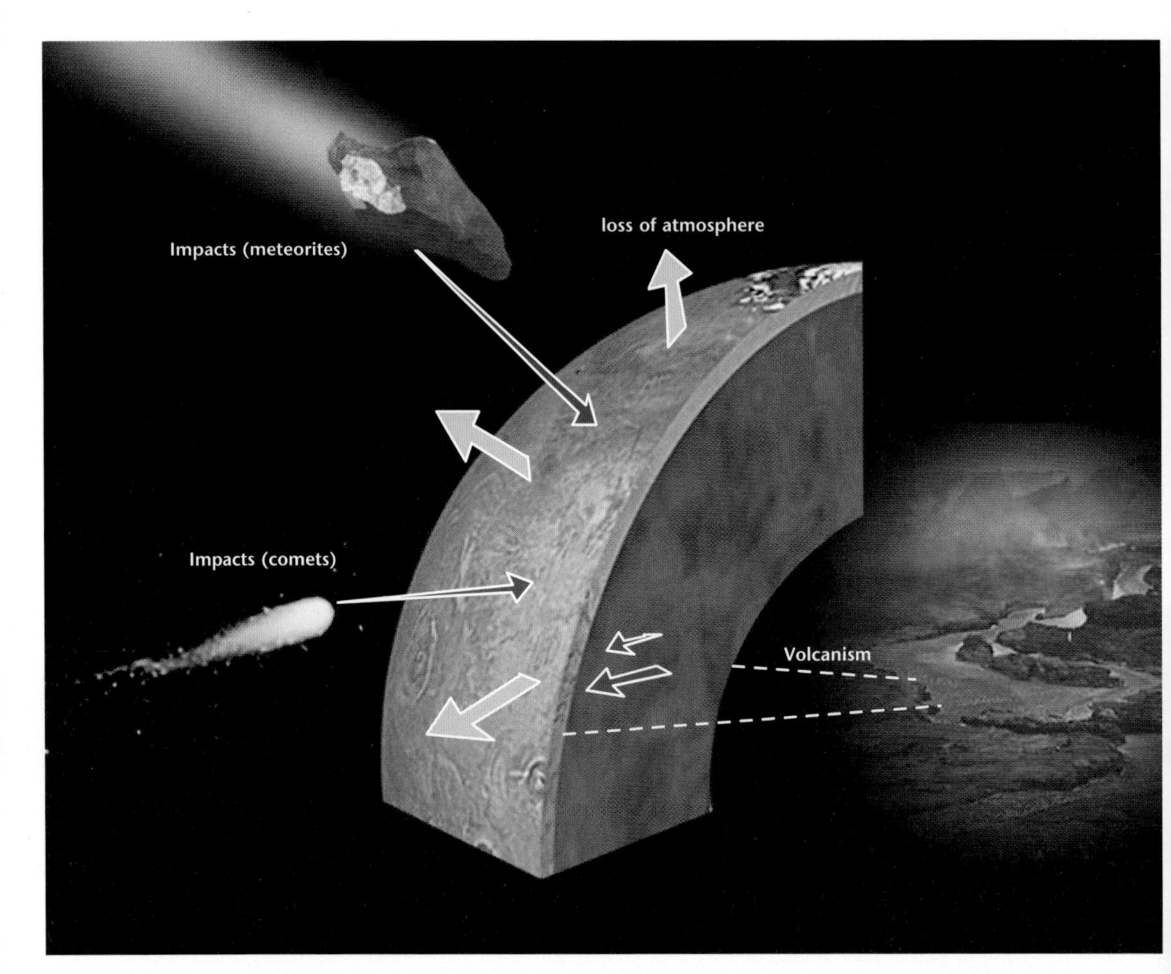

The origin of Mars' water and atmosphere. Most of Mars' water and carbon dioxide was released into the atmosphere during accretion, having been trapped within chondrites. Later, more was released during volcanic episodes. According to Roger Phillips, of the University of Washington, the creation of the Tharsis bulge and its volcanoes would have involved some 300 million cubic kilometres of lava, sufficient to liberate enough water to form a global ocean 120 metres deep and an atmosphere 50 percent denser than Earth's. After the accretionary phase, impacts of water-bearing comets and meteorites would also have contributed to the formation of the primitive ocean and atmosphere. (Courtesy NASA, JSC, JPL, and Calvin J. Hamilton.)

To discern the original composition of the planetary atmospheres, planetologists need to put in a lot of detective work: the compositions have been vastly changed by chemical and physical interactions of the volatile elements, the solid crust and the liquid oceans, and by volcanic activity. Looking for atmospheric constituents which have persisted unchanged for four billion years requires, paradoxically, that those elements which form today's atmospheres have to be left out of the analytical equation. For example, oxygen, nitrogen and carbon form carbonates and nitrates which are incorporated into the crust; deuterium and hydrogen may be products of the decomposition of water in oceans and ice deposits. There remain only the rare gases, characteristically chemically inactive (hence their alternative name of 'inert' gases), and more particularly their isotopes (atoms of the same chemical element which differ both in the number of neutrons in their nuclei, and in their masses).

In 1976 the Viking probes took measurements, indicating that these rare gases make up just over 1.6 percent of Mars' atmosphere. Most of this is argon-40, an isotope produced by the radioactive disintegration of potassium-40 in the rocks, and there are some traces (less than 5 parts per million) of neon, krypton and xenon.

Describing the origin and history of the Martian atmosphere is made even more difficult by the planet's low gravity, lack of a magnetic field (see pages 42–44) and still unknown history of volcanic activity. These two factors are responsible for the loss of most of the primitive atmosphere, which was stripped away as a result of meteoritic impacts and the action of the solar wind.

As for Mars' initial quantity of water, it may well have been greater than that in the Earth's oceans. We know now that after an early phase of accretion when part of the core and mantle were formed, the process continued with the arrival of planetesimals rich in volatile elements and water (forming as much as 10 percent of their mass). Analyses of meteorites originating on Mars seem to indicate that half of the planet is composed of such material. Most of the water has certainly disappeared, but some remains in the crust in the form of ice (see Parts 2 and 3), and in the planet's mantle, although quite how much of it now exists is a matter of debate.

7 Meteoritic bombardment

An unimaginably violent rain of rocks upon early Mars

During the first 700 million years of their existence, the planets of the solar system underwent a phase of intense bombardment by asteroids and small bodies which had not participated in the process of planetary accretion, and had been forced out of their orbits by the gravitational influence of giant planets. The climactic epoch of this bombardment occurred about 3.8 billion years ago, as these giant planets rapidly and finally migrated. Mars, unlike planet Earth, still bears the scars of that episode across more than half of its surface. Most of Mars' southern hemisphere is still covered with craters, of diameters from tens of metres to hundreds of kilometres.

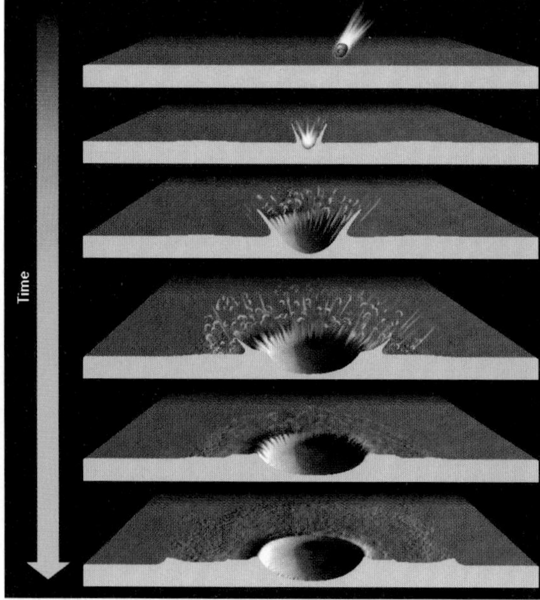

A bowl-shaped crater 2.5 km in diameter, to the north-west of Elysium (illuminated from the right; MOC image). (Courtesy NASA, JPL, Malin Space Science Systems.) Just as a pebble thrown into water can produce only circular ripples, most impacts leave round craters, even if the impactor does not fall vertically. There are, let it be said, a few elliptical craters, corresponding to 'grazing' impacts where the impactors approached at angles below 15°.

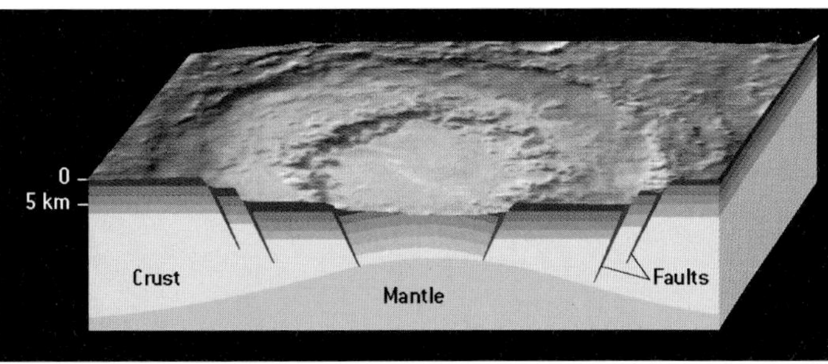

Lowell Crater (left), 201 kilometres in diameter, is situated in the west of the Argyre Basin. In the illustration, illumination is from the left (MGS image). (Courtesy NASA, JPL, Malin Space Science Systems.) Craters like this one, more than 100 kilometres wide, result from extremely violent impacts and often have multiple ring structures. The crust, locally reduced by the loss of the ejecta, undergoes a slight deformation which readjusts the sub-surface features. These readjustments produce numerous circular faults within the crater (see diagram above), which facilitate the rising of lava with subsequent eruptions or lava outflows, as has happened on the Moon. In the latter case, the floors of the craters are smooth and flat.

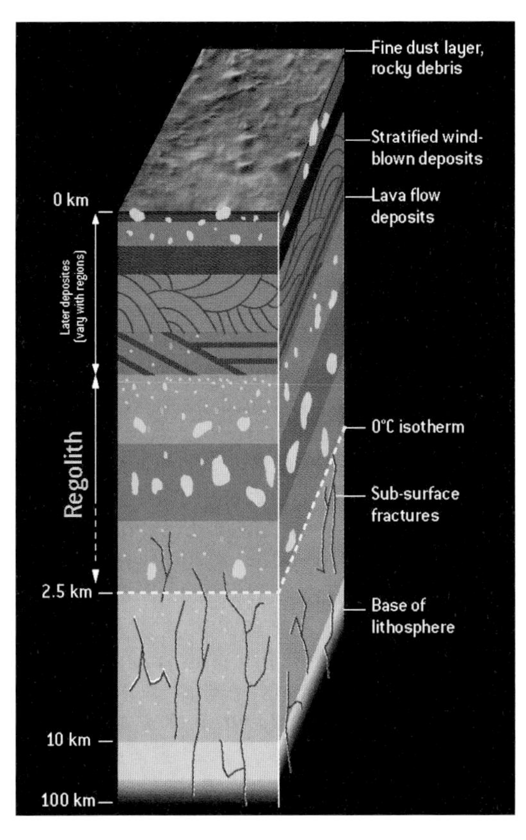

The formation of Mars' sub-surface. Impacts upon Mars during the first billion years of its history formed a 'regolith' (an accumulation of rocky debris of various sizes, the 'fallout' from impacts); this regolith is about 2 kilometres deep, and is uniformly spread across the whole surface of Mars. Therefore, the upper layer of the Martian sub-surface is relatively 'porous', encouraging the retention of water and ice and explaining the presence of a permanently frozen layer beneath the surface (see pages 106–108). Deeper down, the sub-surface material is fractured. Here, the pressure from layers above has probably lessened the volume of the voids between the ejecta debris. Below 10 kilometres beneath the surface, the sub-surface material would probably have no spaces in it at all.

Topography of the high plateau of the southern hemisphere reveals a heavily cratered surface. At the centre is the deepest impact basin in the solar system, Hellas Planitia. It is 11 000 metres deep and 2500 kilometres across. Hellas Planitia is thought to be evidence of a collision between Mars and an object about 450 kilometres in diameter, more than 4 billion years ago. Such cataclysmic impacts could have taken place only at the very earliest stage of the history of Mars. Most of the big impact basins such as Hellas and Isidis exhibit at their peripheries multiple concentric rings of ejecta, several hundred kilometres wide. Data from MOLA, Mars Global Surveyor. (Courtesy NASA, MOLA Science Team.)

The formation of an impact crater

A body just a few hundred metres across, striking the surface of Mars at a velocity of 15 km/s, would produce a release of energy equivalent to several hundred megatons (the value for the Hiroshima bomb was 13 kilotons). Such an impact can eject fragments of the planet's surface, as well as part of the atmosphere, into space; what remains of the atmosphere may be darkened by a rain of debris and clouds of dust.

At the time and location of the impact, pressures several thousand times greater than that of the terrestrial atmosphere, and temperatures of several thousand degrees, cause material to be liquefied, vitrified and even vaporised. Sub-surface material is massively fractured. The compression wave created by the impact is followed by a stress relaxation wave involving the violent decompression of the underlying rocks. This process of excavation expels material at the periphery of the crater, just a few hundredths of a second after impact. Within the cavity, a (slower) phenomenon now occurs as material is readjusted, and in the hours, or even days, following the initial collision, a central peak or ring forms. The ejecta (material thrown out of the crater at the moment of impact), some of it in molten form, lies in a radial pattern around the crater.

Impact craters exhibit a morphology which varies according to size: the smallest craters (less than 20 km across) are bowl-shaped, while the floors of larger ones appear filled-in and sometimes have peaks or rings at their centres.

8 The enigma of the 'north-south dissymmetry'

Why does the high plateau of the south dominate the plains of the north?

A particular feature of Mars is the geographical dissymmetry between the south and the north, characterised by an average difference in altitude of 5 kilometres and an unequal distribution of geological formations. The high plateau of the southern hemisphere is ancient and heavily cratered, while the surfaces of the northern plains are more recent and have comparatively few craters. Between the two areas, there is a transitional zone marked by innumerable groups of buttes (isolated hills with steep sides and flat tops), evidence of intense erosion of the high plateau. The edge of the plateau itself is in fact not well defined, and it peters out with small buttes at some 45° N. The cause of this dissymmetry remains a mystery. Two avenues of research are currently being pursued, one invoking an internal geological origin, and the other an external origin linked to cataclysmic impacts.

The 'internal' hypothesis

Could the dissymmetry be a result of asymmetrical convection within the mantle, leading to a difference in thickness across Mars' crust? The deviations in the paths of orbiting spacecraft allowed scientists to measure variations in the planet's gravitational field. These measurements tell us about the thickness of the crust and the presence of any sub-surface anomalies due to convection in the mantle. What were the results of the study? Variations in the planet's gravitational field revealed that the crust was 40 kilometres thick in the north, and 70 kilometres thick in the south, but this variation seems not to have been caused by asymmetrical convection within the mantle: along the transitional zone between the plateau and the plains, the crust does not exhibit the expected variations in thickness.

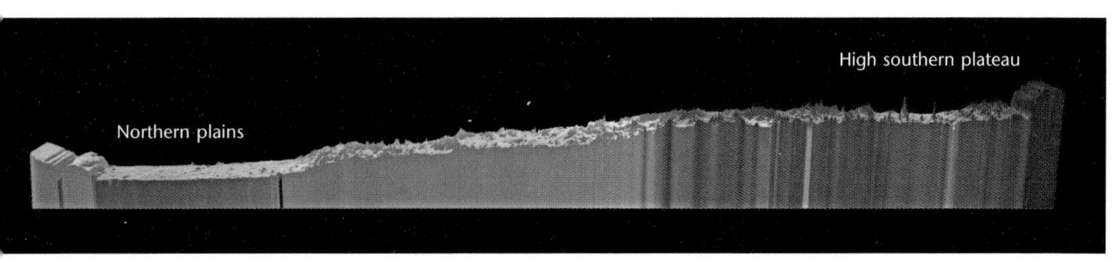

Topographic section showing progressive diminution in altitude from south (+1 to +2 km) towards north (–4 km). Data from MOLA (MGS) at longitude 0° E. (Courtesy NASA, MOLA Science Team.)

The 'external' hypothesis

The crust of the northern hemisphere may have been thinned by vast numbers of meteoritic impacts. Studies of gravitational anomalies, and recent images provided by sub-surface sounding radars suggest the presence of impact features buried beneath deep sedimentary layers. This cratered surface, now hidden by sediments, was probably formed at the same time as the southern plateau. Later, it was to be covered to a depth of several kilometres by mighty deposits of material, products of the erosion of the crater-strewn high plateau. What remains to be understood is why the greatest impacts should have been visited preferentially upon the northern hemisphere.

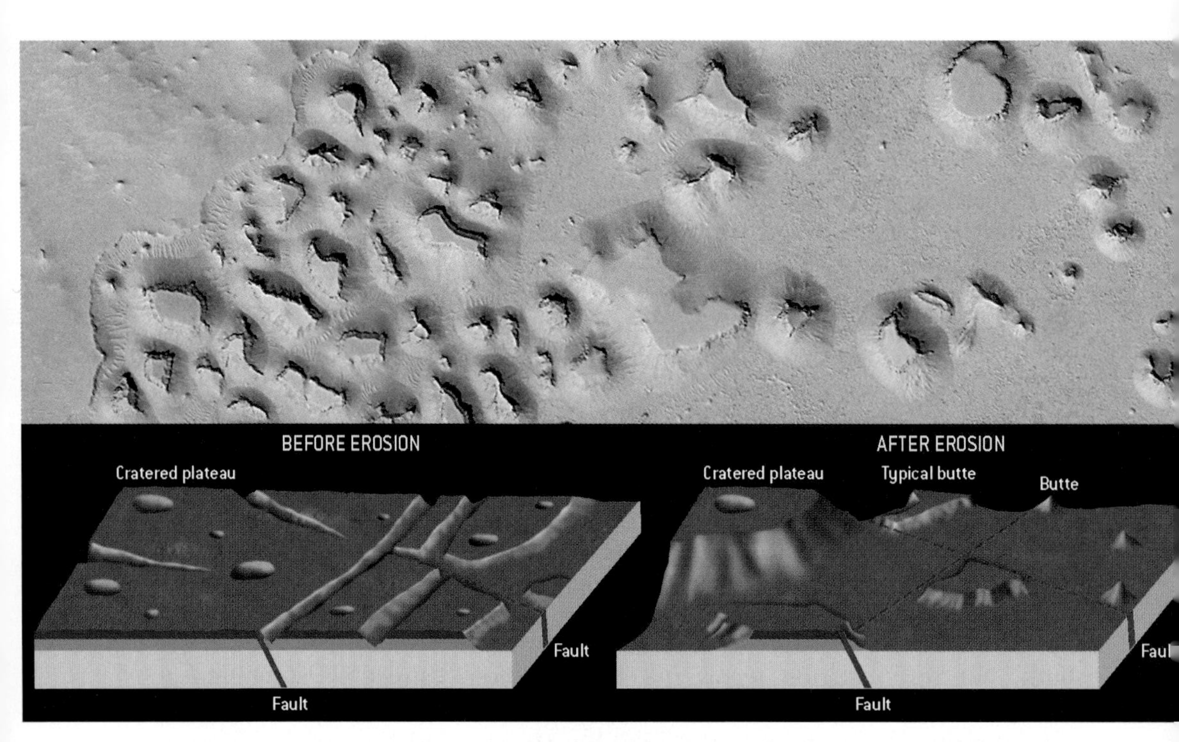

A rugged frontier between the two hemispheres. The transitional zone between the southern cratered plateau and the lower, northern plains is intensely eroded and chaotic along faults and fractures. The upper image is 3 kilometres from top to bottom, with illumination from the left (MGS image). (Courtesy NASA, JPL, Malin Space Science Systems.) The diagram below the image explains the progressive erosion of the cratered plateau: the rectilinear aspect of the valleys and the rectangular aspect of the buttes (hills) are the results of delimitation by faults before the erosion.

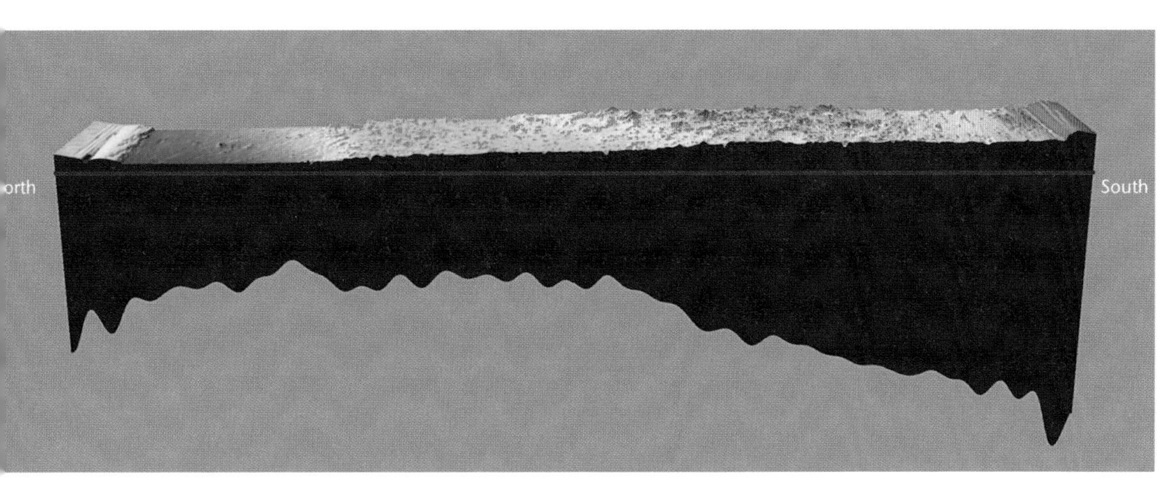

The variation in depth of the Martian crust is visible in this topographical section. On average, the crust beneath the plains of the northern hemisphere is 40 kilometres deep, and that beneath the cratered southern plateau is 70 kilometres deep. Data: MOLA, MGS. (Courtesy NASA, MOLA Science Team, GSFC, Scientific Visualization Studio.)

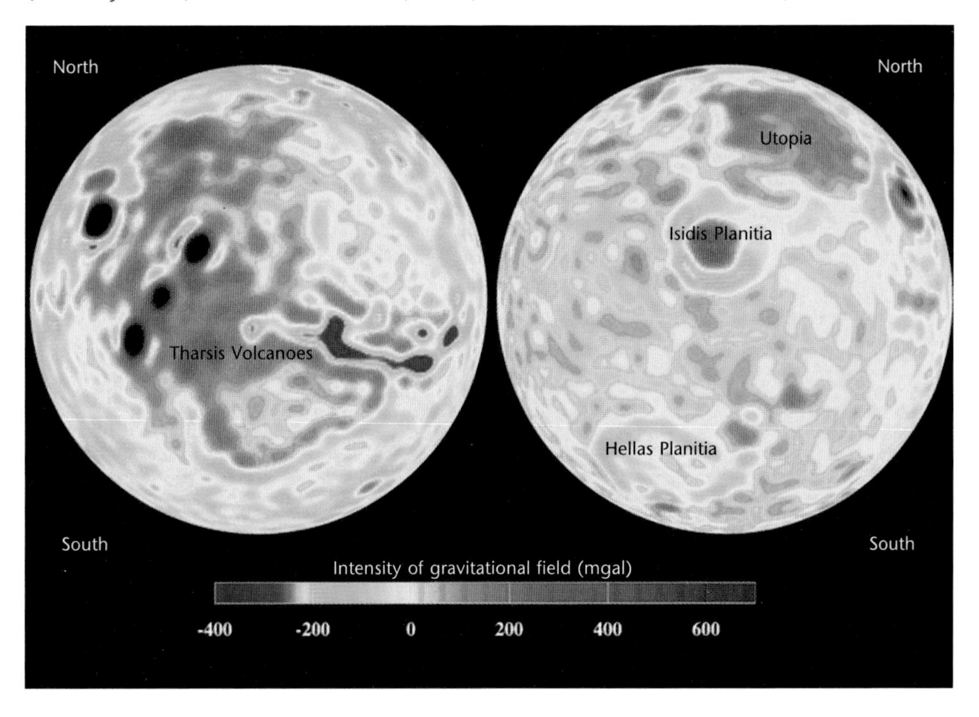

Chart of anomalies in the Martian gravitational field (Mars Global Surveyor). The chart indicates the positions of irregularities in the distribution of mass in the crust and reveals the presence, in the north, of ancient impact craters now hidden beneath sediments (such as the Utopia Basin, comparable to the Hellas region). One mgal is about one millionth of the Earth's gravitational field strength. (Courtesy NASA, MGS Radio Science Team.)

9 The primordial dynamo and magnetic field

Vestiges of an ancient magnetic field stronger than Earth's?

In 1997, Mars Global Surveyor (MGS) became the first spacecraft to pass over Mars at very low altitudes. One of its mission aims was to measure the magnetic field of the ionosphere, the layer of the atmosphere ionised by ultraviolet

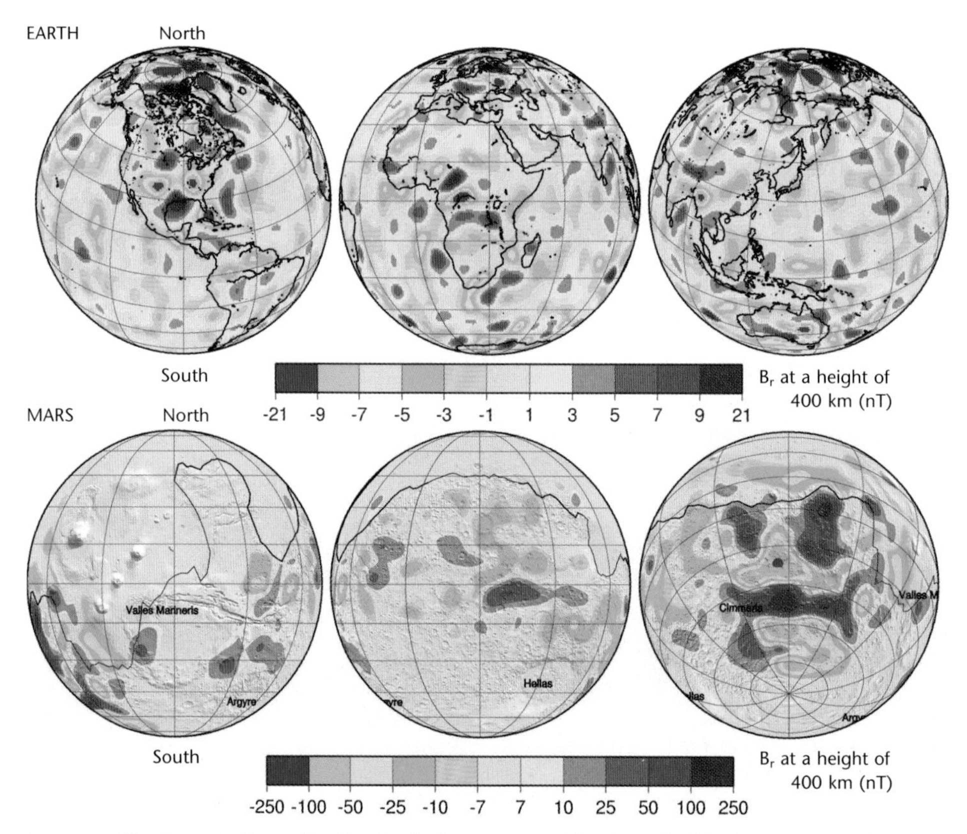

A magnetised crust. Mars, like the Earth, has a magnetised crust with strong magnetic anomalies (B_r). These anomalies, measured in nanoTesla units, are almost exclusively beneath the ancient terrains of the southern hemisphere. The fields that have been measured are ten times more intense than their strongest terrestrial counterparts, measured near Kursk, in Russia. In the Martian atmosphere, these fields locally create small magnetic aurorae, discovered by the SPICAM spectrometer aboard the Mars Express probe. How can such intensities be explained? On Mars, the crust is twice as near to the core-mantle boundary as it is on Earth: the magnetic field created by the ancient Martian dynamo may therefore have been much more influential at the surface in the case of Mars (nT = nanoTesla). (Courtesy Michael Purucker, NASA.)

radiation and the solar wind. As MGS encountered Mars, the magnetometer was activated. As had been expected, the magnetic field showed a maximum as the probe moved through the ionosphere, but, to everybody's surprise, values increased again during some of the low passes above the most ancient terrains on the planet.

MGS had detected the remnants of an ancient and once powerful magnetic field, capable of magnetising the surface; it had previously been thought that Mars had no such field. Planetologists now faced the questions of how it came to be, and why it disappeared.

Magnetic fields in terrestrial-type planets are created by movements within the iron core (see pages 24–26), which is fluid and a good conductor of electricity. The exact mechanism producing such fields is not completely understood. It is

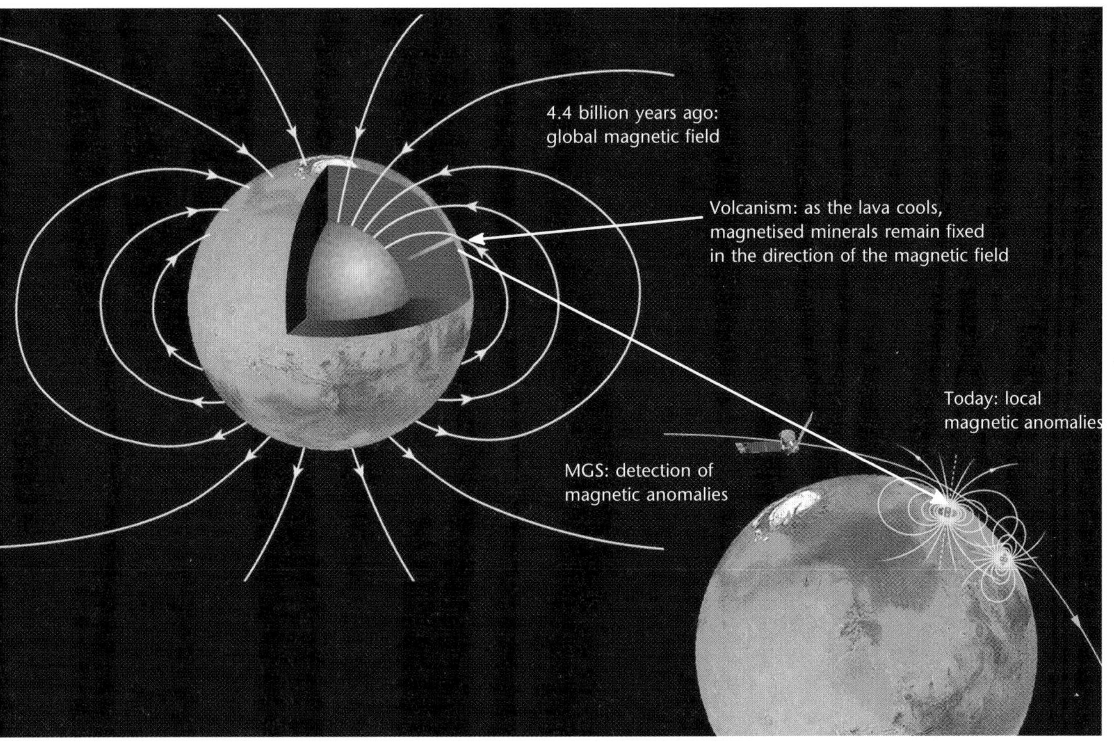

Magnetisation of the surface. The magmatic rocks constituting the Martian crust are often magnetic. Crystals of certain minerals such as magnetite behave like small magnets. With the rock still in the molten state, they spontaneously align themselves in the direction of the planet's magnetic field. When the rock solidifies, these magnetised minerals remain fixed in this direction, even if the planetary magnetic field disappears. The field they produce is therefore a clue to the original field which brought them into position. A magnetic field must have existed in the magmatic rocks which cooled to form the ancient crust of the southern hemisphere. The more recent crust of the northern hemisphere is not magnetised. The magnetic field had already disappeared at the time of its formation. (Courtesy NASA, JPL, USGS.)

thought that convection within the liquid core, with hot iron ascending from the centre towards the boundary with the mantle, there to cool and descend again, generates an electric current. This in its turn produces a magnetic field: the 'dynamo' effect. How does the electric current come about? Apparently, 'randomly' created micro-magnetic fields within this medium generate the electric current, and in its turn a global magnetic field.

In the case of Mars, there existed, just after the accretion stage 4.45 billion years ago, a liquid core hot enough to sustain convection and thereby generate a magnetic field. It seems likely that the core then cooled very rapidly, and some hundreds of millions of years later (500 million at most), the residual heat was insufficient to trigger convection. Conduction (see pages 88–90) then came into its own, and, with the cessation of convection, the Martian 'dynamo' switched itself off.

Today, Mars' crust retains a lingering magnetic 'memory' of the active dynamo. Only by dating samples from the cratered terrains of the south will scientists be able to estimate just when the dynamo ceased working.

10 The moons of Mars: captured asteroids?

Phobos and Deimos: held in a gravitational trap

In Martian skies, two little natural satellites pursue their paths in opposite directions. The larger of the two, Phobos, measures 26.8 × 22.4 × 18.4 km, rising in the west twice a day. Deimos (15.0 × 12.2 × 10.4 km) rises in the east once every five and a half days. Seen from Mars' surface, their apparent diameters are respectively about three times and seventeen times smaller than that of our Moon, as seen from Earth.

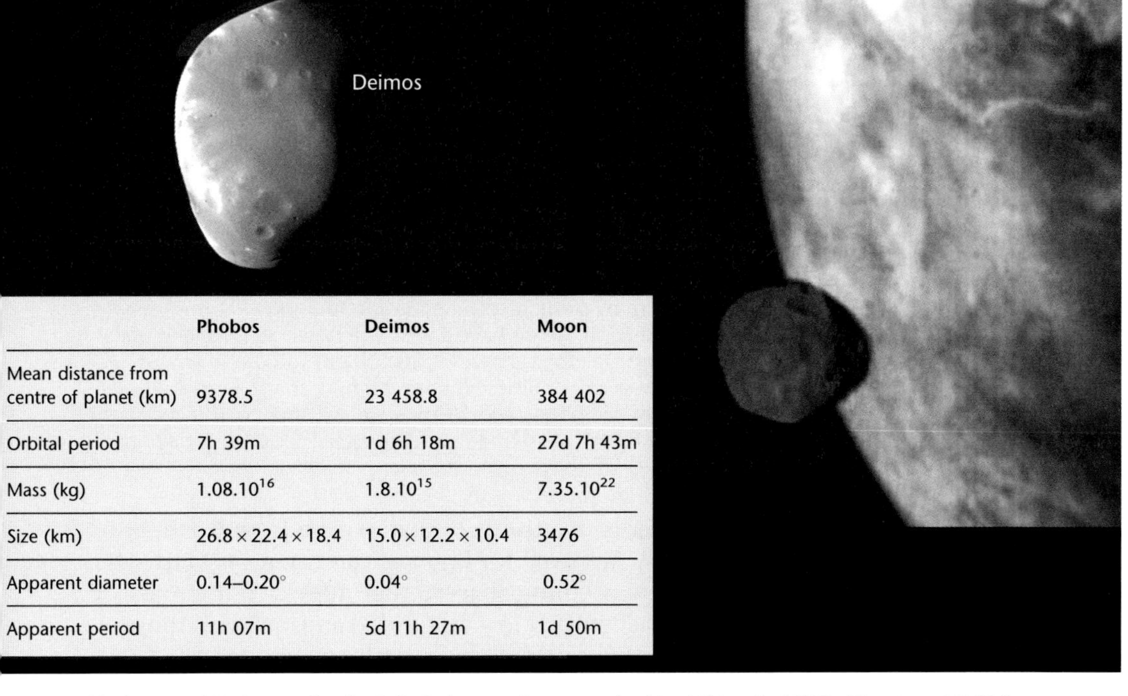

	Phobos	Deimos	Moon
Mean distance from centre of planet (km)	9378.5	23 458.8	384 402
Orbital period	7h 39m	1d 6h 18m	27d 7h 43m
Mass (kg)	$1.08.10^{16}$	$1.8.10^{15}$	$7.35.10^{22}$
Size (km)	26.8 × 22.4 × 18.4	15.0 × 12.2 × 10.4	3476
Apparent diameter	0.14–0.20°	0.04°	0.52°
Apparent period	11h 07m	5d 11h 27m	1d 50m

Phobos and Deimos. On the left, Deimos, photographed by Viking in 1977. (Courtesy NASA.) On the right, Phobos, near to Mars, photographed by the Soviet probe Phobos 2 in 1989 as it approached the planet. (Courtesy Space Research Institute, Russian Academy of Sciences.) Lost prematurely, the Phobos 2 probe was due to study this satellite in detail. It was to put a module on the surface of Phobos, which would have moved around in stages, benefiting from the low gravity. The Russian space agency is now planning a new mission to Phobos, 'Phobos Grunt', this time to return Phobos samples back to Earth in the next decade.

Phobos, photographed from less than 200 kilometres by the camera on Mars Express, on 22 August 2004. The crater Stickney, 10 km by 6 km, is seen on the left. Although Phobos' gravity is a thousand times weaker than that of Earth, debris have 'fallen' into the crater, leaving traces upon its slopes. Gravity on Phobos and Deimos is however insufficient to compress and compact materials: from 10-20% of the volume of these moons is empty space. Even if we take these values into account, the densities of Phobos and Deimos are still small compared with those of other asteroids, a fact that intrigues planetologists. (Courtesy ESA, DLR, FU Berlin (G. Neukum).)

As the Sun's apparent diameter is two-thirds of that as seen from Earth, only very partial solar eclipses, already observed for Phobos, can occur on Mars. The sizes and shapes of Phobos and Deimos suggest that they are asteroids, a hypothesis corroborated by the fact of their low densities compared with those of the Moon or Mars. Phobos' density is almost twice that of water, and the value for Deimos is 1.7, while for the Moon it is 3.34, and for Mars, 3.93. Also, the satellites' surfaces absorb nearly 95% of incident sunlight. They share this property with certain asteroids orbiting the Sun between Mars and Jupiter, at distances between 3 and 3.5 times that of the Earth from the Sun (i.e. 450–525 million kilometres).

If Phobos and Deimos are indeed asteroids, we can speculate as to how and

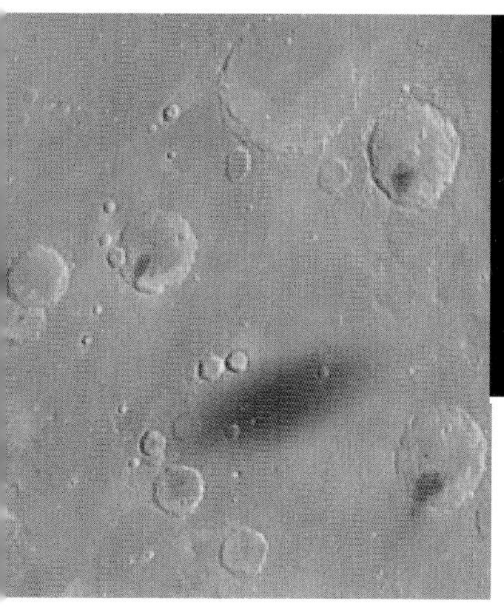

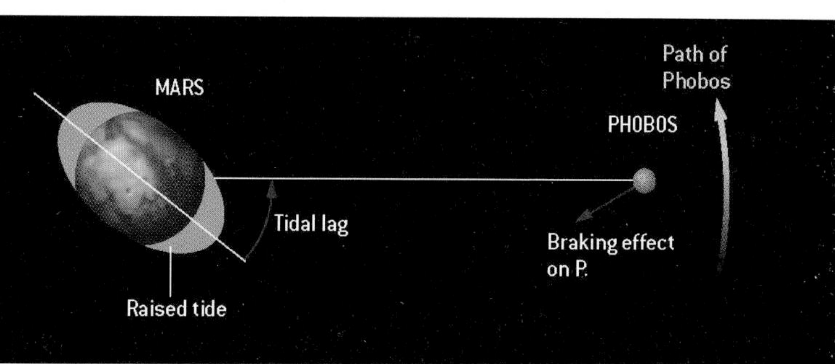

The shadow of Phobos on the surface of Mars (left), captured by Mars Global Surveyor. (Courtesy NASA, JPL, Malin Space Science Systems.) The shadow is not the only phenomenon produced by Phobos on the planet: even though its mass is extremely small, Phobos 'deforms' Mars by creating a solid tide, a raising of the surface by one millimetre, thirty times less than the effect produced by the Moon on the Earth. Because of the viscosity of Mars' mantle, this 'deformation' occurs after a certain time lag. The 'high tide' always occurs after the passage of the satellite, which creates a braking effect on it. As a result, Phobos will crash into the Martian surface in about 40 million years' time. The effect is reversed in the case of Deimos. Like our Moon, it is slowly moving away from the planet.

when they assumed their present positions. Mars must have captured them in the same way as Jupiter, Saturn and Neptune seem to have captured some of their satellites. This does not really answer the question, however: just as space probes need to fire their engines to achieve a braking effect and go into orbit, 'something' has to slow an asteroid in order for capture to take place.

Several models have been envisaged. The first, echoing the story of the Earth and its Moon, supposes that a body collided with the planet, sending debris into orbit; the fragments then re-accreted. Unfortunately, this mechanism is unlikely to be applicable to Phobos and Deimos: they are not dense enough. Another model brings in 'aerobraking', with the body being sharply decelerated as it passes through a dense atmosphere. Such an atmosphere existed on Mars at the time of its formation, but was soon swept away by solar activity. If this model reflects reality, then Phobos and Deimos could be the last representatives of the planetesimals from which Mars was made, between 4.5 and 4.3 billion years ago. The question remains open.

A youthful Mars

–4 to –3.5 billion years

An oasis where life might have flourished? Our story continues on a planet astonishingly different from the one we know now. Its geological archive reveals episodes from a young world, frequently bombarded by bodies still orbiting the Sun in large numbers. On this youthful Mars, temperatures are such that liquid water can exist, at least episodically; rivers and lakes erode the surface and sediments accumulate.

At the same time, 3.8 billion years ago, the first micro-organisms were emerging and developing on Earth, where conditions were scarcely more favourable than those on Mars. Could life also have begun on the red planet?

Water and craters on Mars. The crater Fesenkov, 4 billion years ago, as imagined by Kees Veenenbos.

1 Rivers on Mars!

The youthful Mars as revealed by the tracks of rivers long since vanished

In 1972, Mariner 9's discovery of strange valleys resembling dried-up riverbeds was nothing short of astonishing. Soon, similar 'fluvial' features were being found all over the cratered plateau of the southern hemisphere, 3.8 billion years old. However, more recent terrains showed little sign of these valleys. So it seemed that, about 3 billion years ago, Mars had harboured conditions

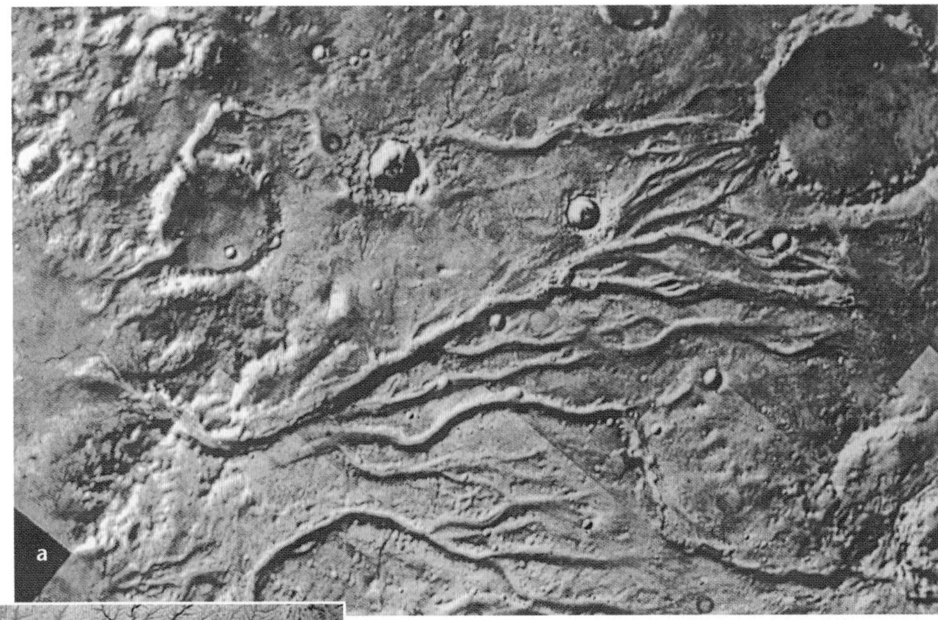

(a) A network of branching valleys, now dried up (Vedra and Maumee Valles, near Chryse Planitia), compared with a network on the Hadramout Plateau **(b)** in southern Yemen, as photographed from the space shuttle Discovery. Both images show an area about 100 kilometres across. Contrary to what we observe on Earth, almost no valley on Mars is narrower than 200 metres, which suggests a different mechanism in the formation of valleys on Mars and Earth. (Courtesy NASA (a) and Earth Sciences & Image Analysis Laboratory, NASA (b).)

Valley networks, outflow channels and gullies on Mars: not to be confused with each other ...

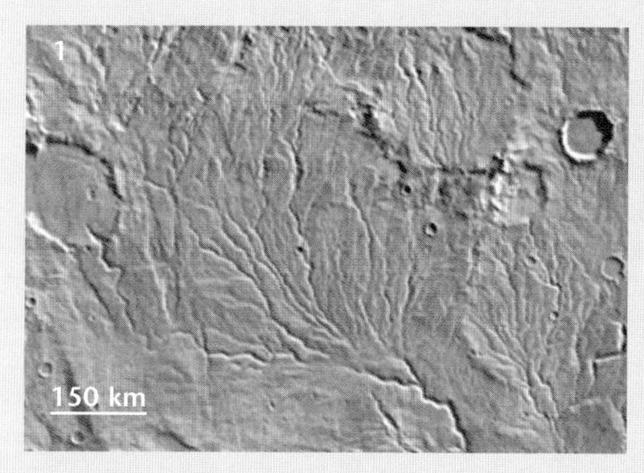

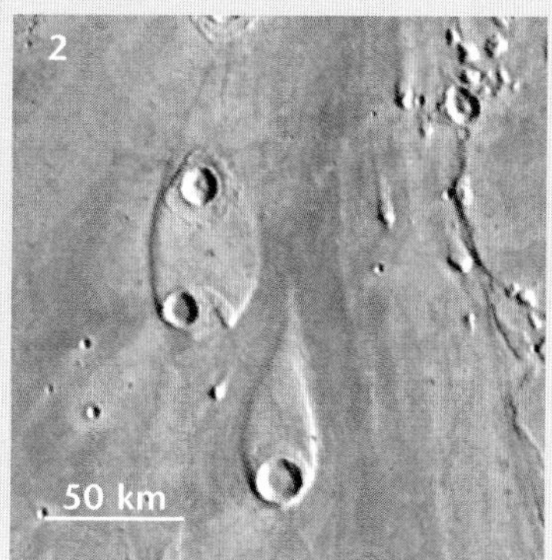

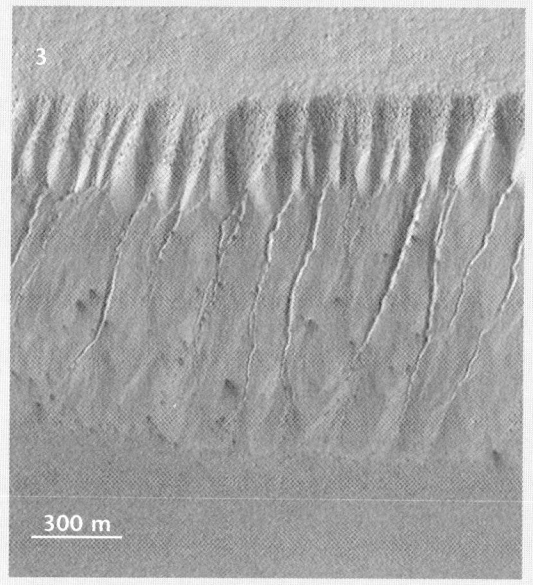

Throughout this book, we discuss very different kinds of flow features created by liquid water.

1. Many dendritic 'fluvial' valleys (some meandering, e.g. Nirgal, Nanedi and Ma'adim) were formed more than 3.5 billion years ago, probably when the climate was more clement than today, as described in this section.
2. Huge outflow channels (see pages 111-113) are more recent, their widths exceeding 100 kilometres. These channels were probably made by violent, catastrophic outflows at various times in the course of Mars' history, although the climate may not have been very different from what is observed today.
3. Small traces of gullies (see pages 176–178), the features some hundreds of metres long, apparently date from a few hundred thousand years ago, or even less: 'yesterday' in geologic terms.

(Images courtesy NASA and Calvin J. Hamilton (1 and 2), and NASA, JPL, Malin Space Science Systems (3).)

favourable to the creation of such fluvial valleys; conditions which have not recurred since. But were we really seeing evidence of rivers?

Most of these valleys, some of which are up to 1000 kilometres long, show a large number of separate branches, with tributaries coming together as in a drainage basin. This strongly tree-like layout suggests that the episode (or episodes) during which liquid flowed lasted a relatively long time. Planetologists concur that the liquid that carved the valleys was indeed water; scenarios involving lava and liquid carbon dioxide have been envisaged, but have proved unconvincing. One reason why it is not easy to compare Martian valley networks with their more recent Earthly counterparts is their great age: time slowly erases certain characteristics. For example, it was long thought that the number of branches was far less than might be found on Earth, but more up-to-date images from a thermal imaging camera have revealed many previously invisible tributaries, in systems whose densities rival those on Earth. Many questions suggest themselves. Is a warm and moist climate absolutely necessary to unleash such flows? Could a cold, Arctic-type climate have had the same effect? Do we need to involve a water cycle, with rain soaking into the surface or running across it, to explain what we see? Whatever the case may be, it is certain that the surface water must have been in a stable, liquid state to carve such valleys, implying that the atmosphere in those days was denser than it is now, at least episodically.

2 The source of Mars' rivers

The uncertain origin of the ancient rivers of Mars

Ancient valley networks on Mars may resemble similar features on Earth, but there are differences. Still actively debated are questions such as: do we really need to invoke terrestrial-type precipitation and flows in Mars' past? It could be that some fluvial valleys on Mars might owe their origin to hydrothermal circulation, or even underground water gushing from springs.

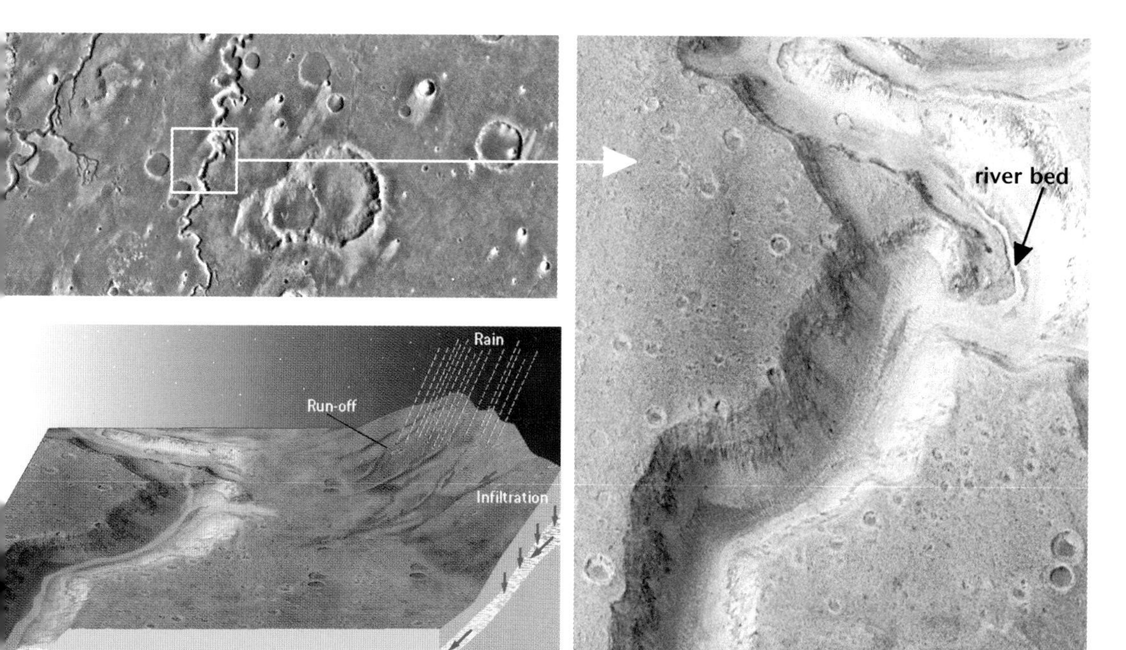

Nanedi Vallis, 2 or 3 kilometres wide and 1000 kilometres long. The close-up on the right shows an area 10 km by 28 km, with illumination from the left. This is an exceptional image, showing as it does a 'riverbed' about 100 metres across, confirming the fluvial origin of the valley. Almost everywhere else, the valley floors have been covered with dunes and sediments. It is easy to discern terraces, evidence of a progressive incision of the watercourse. Bottom left: a reconstituted view. (Images courtesy NASA, JPL, Malin Space Science Systems.)

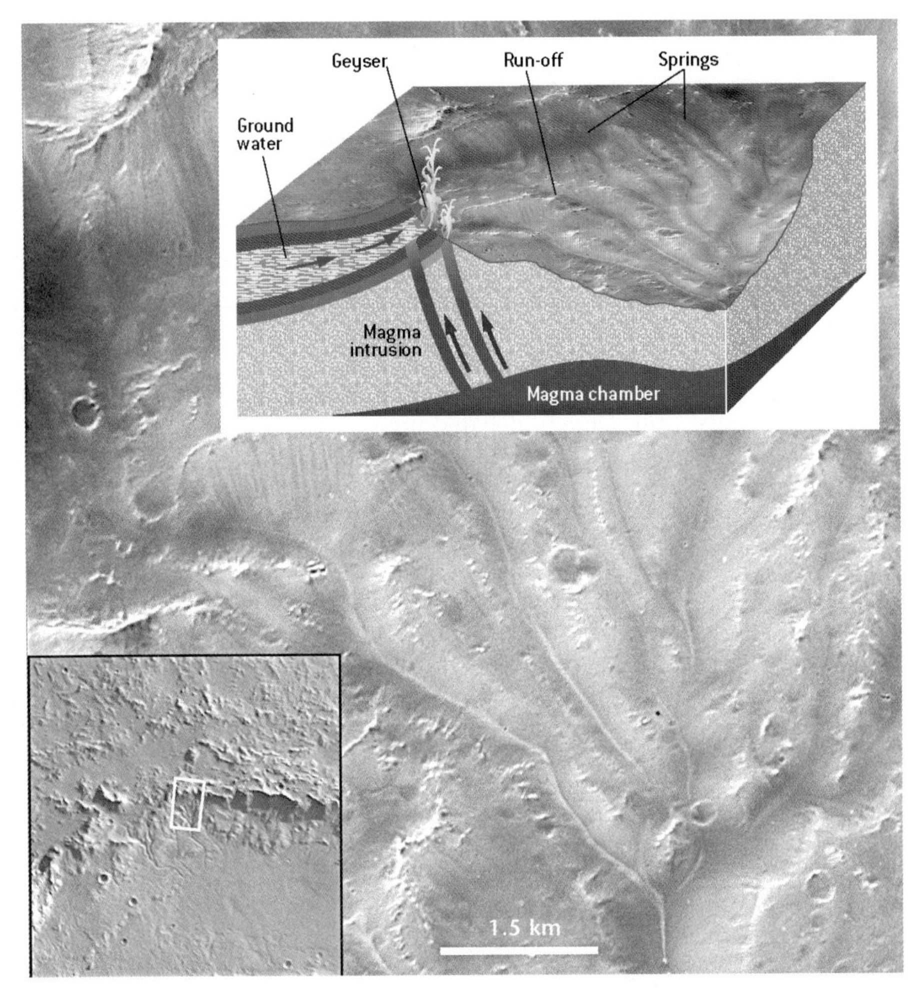

Channels suggesting underground water sources in Bakhuysen crater. The sources are confined to the internal wall of the crater, and their formation could have involved underground reservoirs of water, or some kind of hydrothermal circulation. (Courtesy NASA, JPL, Malin Space Science Systems.)

Hydrothermal sources

A certain number of branching valleys originate on the slopes of volcanoes, or impact craters such as the crater Bakhuysen, seen in the photo above. One explanation of this phenomenon might be that, near a volcano or after an impact, rising magma heats sub-surface water and creates a convection cell (hydrothermal circulation). The heated water comes to the surface to form springs, and rivers are born and supplied. Lava may well up within circular fissures, vaporising water below the surface and giving rise to hot springs and even geysers.

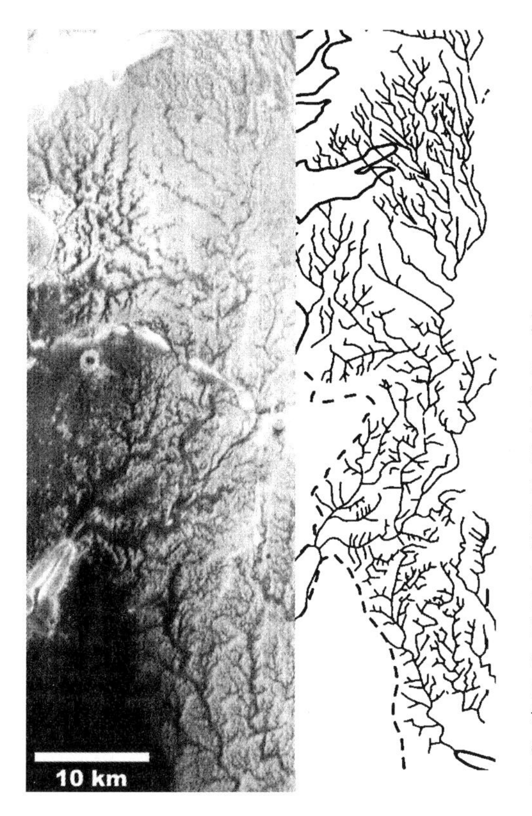

A hydrographic network revealed by the Themis infrared camera on Mars Odyssey (far left). (Courtesy THEMIS Public Data Release, Mars Space Flight Facility, Arizona State University, Nov. 2005.). This network is in the region of Echus Chasma, to the north of Valles Marineris. The great sensitivity of the camera, combined with the nature of the soil and topography, reveal on this nocturnal infrared image numerous branches which would be invisible to a normal camera. The density of the branches is comparable to what might be found on Earth. This implies formation by flowing water, originating in atmospheric precipitation (rain or snow?). Diagram after N. Mangold *et al.*, 2004.

Outgushing of sub-surface water

There is a particular type of branching valley, long, deep-cut and meandering, on the cratered southern plateau. Examples are Nirgal Vallis, Ma'adim Vallis and Nanedi Vallis (see page 53). Unlike the networks previously described (pages 50–52), these features have few tributaries. Surface water run-off has been suggested as a contributory factor, but fails to explain the great depth (of several hundred metres) to which the valleys have been carved. One possible alternative is that the flanks of the valleys have been eroded by subsurface water flows, a process known as 'sapping', rather than by water moving over the surface.

The role of rain

More and more researchers are coming to the conclusion that rain or snow have played a major, or at least episodic, part in the vast process of erosion which marks the most ancient terrains on Mars (see pages 62–64). Even in those places whose morphology invites explanations involving ancient underground water reservoirs, precipitation in former times would probably have been necessary to feed those reservoirs.

3 The mystery of the lakes

Recent observations suggest that some sediments were formed in lakes

There are dozens of impact craters on Mars which may once have contained lakes. 3.8 billion years ago, in an environment favouring the presence of liquid water, the bowl-like shapes of craters could well have encouraged the formation of areas of water, and the accumulation of sediments. For a long time, this was just a theory. Then, in 1999, images from Mars Global Surveyor found all-too-obvious evidence of such features. The floors of certain craters showed finely

Sedimentary deposits on the floor of a crater in the Arabia Terra region. Probably laid down in a lake environment, these deposits would once have covered the whole area. The rock has been progressively eroded after deposition, leaving only a few terraced hills amid the sand dunes. (Courtesy NASA, JPL, Malin Space Science Systems.)

The remains of ancient alluvial material in a delta at the end of a Martian valley, on the floor of an ancient lake which has now disappeared. This exceptional site was discovered by the camera of Mars Global Surveyor in 2002, in Eberswalde Crater, just north of Holden Crater, 1000 kilometres north of the Argyre Basin. The sediments have, over time, been covered up and then re-exposed. The existence of such characteristic deposits proves that a flow of liquid water capable of carrying alluvial material was able to exist on Mars for very long periods. Other stratified sedimentary deposits on the planet probably have a similar origin. (Courtesy, NASA, JPL, Malin Space Science Systems.)

stratified layers, both dark and light, usually located in their central regions. On Earth, this kind of sedimentary deposition is often associated with lakes. Scientists concluded that lakes had once existed on Mars.

The sediments found in Martian craters might have been laid down in lakes fed by rivers: this would explain the delta-like structures exhibited by some deposits at the centres of craters. The frequent presence of one or more channels cutting through the rims of the craters (see previous page) is evidence of fluvial activity and suggests how water may have got into the crater. The dating method involving the counting of small craters (see 'Chronology of Mars', pages 12–14) confirms that these deposits are very old, about the same age as the fluvial valley networks of the high plateau of the south (pages 39–41). In many cases, sediments buried for a very long period seem to have been uncovered at a later date by erosion, a phenomenon also seen, but on a lesser scale, on Earth.

In spite of all these convergent clues, it is not always proven that the sediments in question originated in lakes. In some cases a deposition scenario can be imagined which does not involve water at all. For example, the strata could be the result of volcanic activity, depositing ash. Winds may also have brought dust to these locations. The latter explanation would mean that the sediments are stratified dunes, whose deposition would vary in a cyclic fashion in tune with climatic changes (see Part 4). This hypothesis is more likely than it seems at first sight, given the fact that Mars has been subject to wind erosion for the last three billion years. What might the winds have created over such a long period of time? Nothing on planet Earth can offer any comparison.

4 Traces of water

Salts and clay in some sediments provide new clues to Mars' humid past

A small cliff of sedimentary rocks discovered by the rover Opportunity on the flank of an impact crater named Endurance , almost a kilometre away from its landing point. The sediments, composed essentially of sulphates, and probably formed in the presence of liquid water, have been excavated to a depth of tens of metres by the meteoritic impact. (Courtesy NASA, JPL, Cornell.)

Liquid water is capable of much more than just the mechanical action of carving out river valleys and transporting sediments. As has happened on Earth, it may also have changed the chemical composition of rocks on Mars. Laboratory mineralogical analysis of these rocks would be, of course, the best way to investigate this, but it is not yet possible. However, the nature of Martian minerals can be studied at a distance, by comparing the spectral characteristics of rocks (for example, their colour) with those of samples here on Earth. In 1998 the TES spectrometer aboard Mars Global Surveyor detected a very interesting mineral: crystalline hematite. This is an iron oxide, whose formation often requires the presence of liquid water. Now, hematite has been detected in ancient sedimentary regions on Mars, such as Terra Meridiani, south of Arabia Terra. Could there have been a lake there?

An outcrop of sedimentary rocks (photographed in false colour) near the Opportunity landing site in Terra Meridiani. The rock consists mainly of sulphated salts, deposited in fine layers, probably in the presence of liquid water. Tiny mineral spheres (shown in blue), millimetres across, are incorporated into the rock. Some have fallen to the ground and carpeted the surface, as the rock has been eroded away by the wind. The spheres are concretions, or nodules, of hematite (iron oxide) which had already been detected from orbit – their enigmatic presence on Mars was the reason why Opportunity was sent to this site. (Courtesy NASA, JPL, Cornell.)

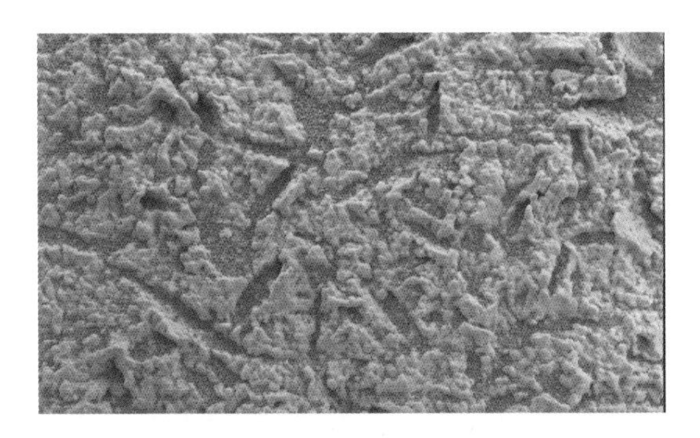

Imprints left by ancient crystals in a rock in Terra Meridiani, through the microscope on Opportunity. Salt crystals would initially have formed after the evaporation of water laden with salts. The cavities, or 'vugs', a few millimetres wide, were left when the salt crystals dissolved as they once again came into contact with water. (Courtesy NASA, JPL, Cornell, USGS.)

A robot geologist

In an attempt to clarify this question, Terra Meridiani was the chosen site for *in situ* exploration by Opportunity, a NASA rover laden with scientific instruments (see pages 201–203). Opportunity landed without a hitch on 25 January 2004 inside a small crater 22 metres across. This was a lucky chance, and the lander's camera revealed a light-coloured, layered outcrop of rocks on the walls of the crater. Opportunity analysed these rocks and the presence of hematite was confirmed. It occurred as millimetre-sized spherical grains, nicknamed 'blue-berries', embedded in the bedrock, like berries in a muffin. The existence and configuration of centimetre-scale layerings that appeared to have preserved the tracks of waves in shallow water suggested deposition by liquid water. Importantly, Opportunity's instruments discovered that the rock itself was composed essentially of hydrated sulphate salts. On Earth, deposits of this nature form at the bottom of shallow lakes, and especially when evaporation

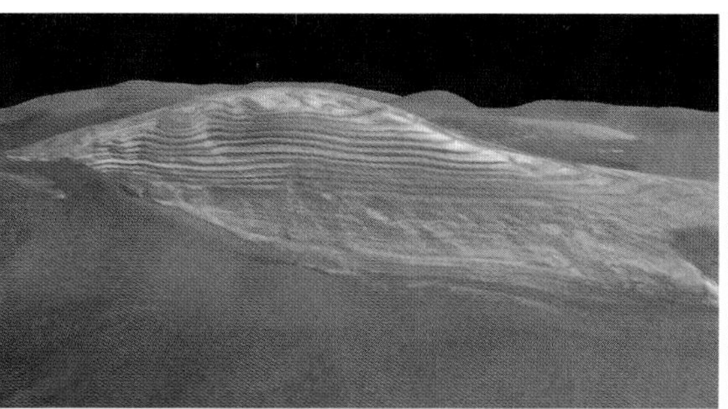

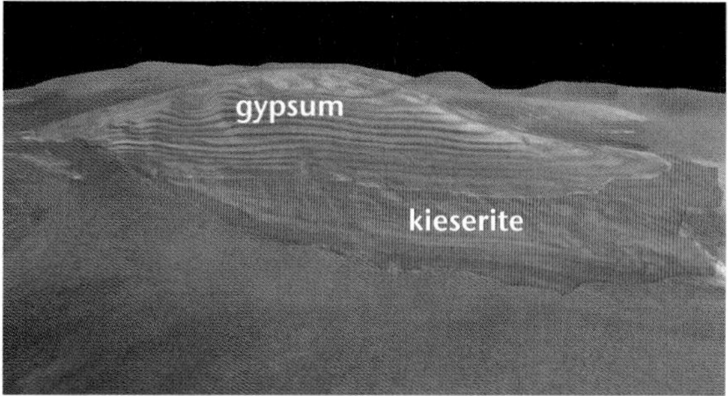

3-D view of a sedimentary deposit on the floor of Juventae Chasma, to the north of Valles Marineris, observed by the HRSC stereo camera on Mars Express. The composition of this mountain, 40 kilometres long and 2500 metres high, was revealed after simultaneous observation by the OMEGA mapping spectrometer. Terraces rich in gypsum (calcium sulphate) seem to have been laid down above an initial deposit of kieserite (magnesium sulphate). The formation of these two minerals usually requires the presence of liquid water. (Courtesy ESA, DLR, FU Berlin (G. Neukum), OMEGA.)

concentrates dissolved minerals. It seemed to geologists that this part of Terra Meridiani had undergone a similar history.

A mineralogical camera

While Opportunity was scrutinising its rocks at the Terra Meridiani site, the Mars Express orbiter (see pages 205–207) was carrying out a global cartographic survey of the mineralogy of Mars with its OMEGA camera. This new kind of instrument is a clever combination of spectroscope and imaging device. It has discovered several sites where sulphate salts are abundant. To the north-east of Opportunity's location, it has revealed a vast zone, hundreds of kilometres wide, where sulphates occur not just on certain slopes, but make up most of the surface. Sulphates have been shown to be the main constituent of many sedimentary deposits on the floor and walls of Valles Marineris, which confirms that liquid water played a part in their formation. However, the story does not end there: OMEGA also discovered clays, which are evidence of another epoch when liquid water flowed in a different environment (see next section).

5 A history of many chapters

Great quantities of water – but only in the very distant past

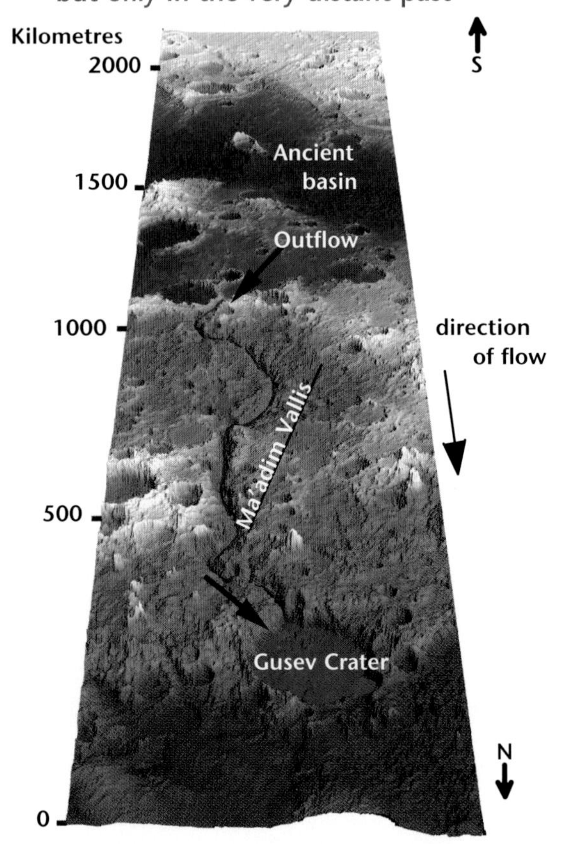

A schematic 3-D view of Ma'adim Vallis between 3 and 4 billion years ago. The valley, several thousand metres deep, runs northward for more than 900 kilometres. It was probably formed in a few months by an outflow, more than 3 billion years ago, of 100 000 cubic kilometres of water from the high plateau of the south onto the plains of the northern hemisphere. The water would have filled the crater Gusev, the destination of the Spirit rover in 2004. Courtesy R. Irwin III, CEPS, NASM/UVA.)

For the first seven hundred million years of its existence, planet Mars probably experienced a number of distinct episodes when liquid water was more or less abundant. The broad picture is difficult to reconstitute, since time has fogged our vision, and each episode has largely erased the traces of its predecessor.

The many faces of Gusev crater

On 4 January 2004, NASA's roving vehicle Spirit touched down successfully on the floor of the large crater Gusev. The rover landed here for a good reason: it was believed that a lake had once lain in this crater. Gusev is situated at the end of Ma'adim Vallis, a long valley cutting through the high plateau of the south for a distance of 900 kilometres. Topographical measurements showed that Ma'adim Vallis originates at a spillway in the divide of a vast drainage basin from which vast quantities of water flowed in the past into Gusev Crater, where considerable sedimentary deposits have been observed. Planetologists, who had high hopes that Spirit might encounter these, were initially very disappointed: for

The floor of Gusev as revealed by Spirit in January 2004: a plain scattered with volcanic rocks. The expected deposits, associated with the ancient lake, have apparently been buried. In the foreground can be seen the deflated airbags used to cushion Spirit's landing. In the distance, 3 kilometres away, are the Columbia Hills, named in memory of the astronauts lost in the space shuttle Columbia on 1 February 2003. Spirit would climb these hills a few months later, finding traces of ancient water. (Courtesy NASA, JPL, Cornell.)

five months, it found little on its 3-kilometre path apart from basaltic volcanic rocks. Lava flows had covered any sedimentary deposits left by any lake in this area. However, on 10 June, Spirit reached the base of a hill, and everything changed. Here were apparently much older, stratified rocks, which seemed to have been in water in the past. It was as if the hill represented an island in the lava flow. Hematite was present (see previous pages), as were various hydrated minerals, like sulphate salts, or goethite, and oxidised iron mineral that cannot form without water. Researchers were still debating, though, whether this modification by water was in fact linked to Ma'adim Vallis: could it perhaps date from a far more ancient epoch, of which few traces now remain?

Clay before sulphates?

The OMEGA mapping spectrometer on Mars Express discovered, in certain specific locations, two types of Martian rocks characteristic of formations originating in the presence of liquid water: sulphates (see above) and clays. These

The rock with 'tentacles'. Observe the shadow of this small rock, discovered by Spirit at the foot of the Columbia Hills. It seems to have sprouted tentacles. On this object, christened the 'Pot of Gold', Spirit found stratified formations rich in sulphur and phosphorus, incorporating tiny millimetre-sized balls composed essentially of hematite. These may be compared to those discovered in Terra Meridiani (see pages 59–60). Geologists suggest that these concretions (nodules) may have come about as a result of geochemical processes requiring water. (Courtesy NASA, JPL, Cornell.)

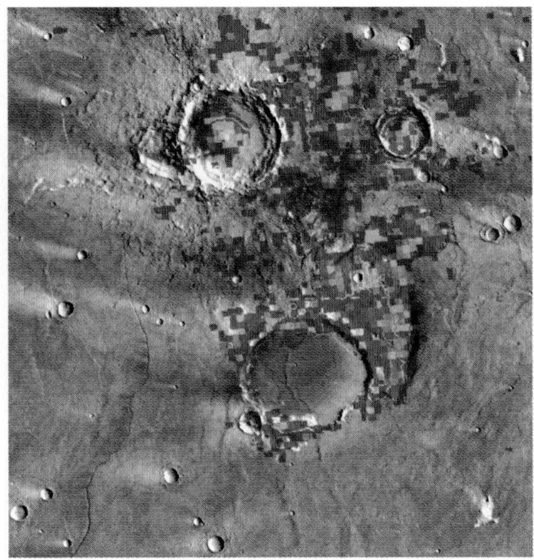

Very ancient buried clays, exposed by meteoritic impacts and wind erosion. In blue on this image, they were detected by the OMEGA mapping spectrometer in the ejecta around craters in the Syrtis Major region. These clays were probably formed by the action of liquid water on volcanic rocks, over tens of thousands of years. (Courtesy NASA, JPL, ASU, OMEGA, Mars Express.)

clays were found in extremely ancient terrains, and especially where wind erosion or a meteorite impact had uncovered long-buried deposits.

We can therefore surmise that Mars experienced, early on, a period when liquid water was probably abundant, and able to soak soils during episodes long enough for clays to form. Later, in a drier and more acidic environment, sulphated salts were more likely to form.

6 In search of a lost ocean

A controversial theory proposes that an ocean once existed in Mars' northern hemisphere

The hypothesis

At the end of the period of intense meteoritic bombardment, 3.8 billion years ago, the topmost kilometre of the Martian sub-surface must have held a quantity of water equivalent to a global ocean 500 metres deep. The observed evidence of former rivers and lakes on the high plateau of the south suggests that the climate, in concert with volcanic and hydrothermal activity, was probably able to maintain a water cycle at the surface. Given these circumstances, vast

Mars under water. These two simulations show what Mars might have looked like if it had been covered by an ocean (shown in blue) to a level 500 metres below the planet's 'zero level' contour as established in the reference system of the MOLA altimeter. The left-hand image shows Hellas Basin under water, and on the right we see the flooded Valles Marineris (see pageS 103–105). Simulations from topographical data of MOLA altimeter, Mars Global Surveyor. (Courtesy NASA, GSFC, Scientific Visualization Studio.)

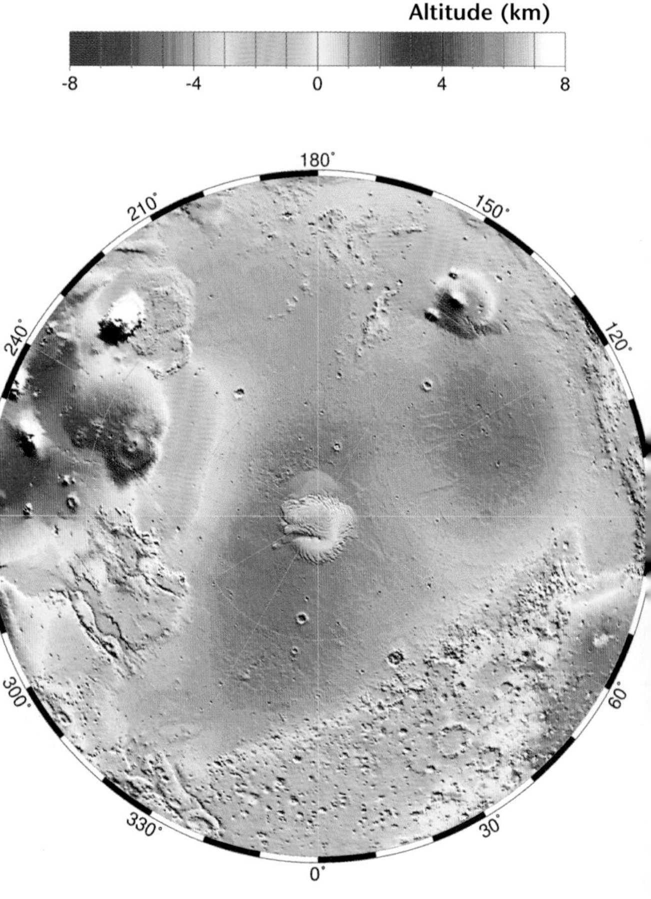

Topographical map centred on the north pole of Mars. This region consists mainly of vast plains: Vastitas Borealis, Chryse, Arcadia, Acidalia and Utopia Planitiae. These plains are lightly cratered, suggesting that their surfaces are more recent than those of the highlands of the southern hemisphere. This northern area may once have held a temporary ocean. This hypothesis is supported by the fact that the surface here is fairly flat and smooth, like the Earth's oceanic abyssal plains. The image above is a view, from the south, of the basin of Utopia Planitia, with vertical scale exaggerated 200 times. We see topographical features, perhaps ancient shorelines, 4350 metres beneath the 'zero level' of the planet. It is estimated that such an ocean would have been 600 metres deep (data: MOLA). (Images courtesy NASA, MOLA Science Team.)

amounts of water and sediments must have flowed towards lower-lying regions, and in particular onto the northern plains. It is therefore possible that an ocean of liquid water or mud existed, if only temporarily, in an extensive area known as 'Oceanus Borealis'. This would have been four times larger than the Arctic Ocean. At that time, Hellas Planitia could have been a veritable inland sea, and the impact basin Argyre could have been connected to the northern plains.

Having formed, this ocean would progressively have disappeared, as evaporation and infiltration into the subsoil took their toll. The climate evolved towards its present glacial state, locking some of the water into the ground in the form of ice. Some planetologists even think it possible that oceans formed and re-formed on a cyclic basis for the first few billion years in Mars' early history.

Are there any traces of this former ocean?

Data from the MOLA laser altimeter aboard Mars Global Surveyor supported the hypothesis of the northern ocean. In the zones corresponding to ancient shorelines, MOLA revealed that altitudes were fairly constant for a distance of some 1000 kilometres. This tilts the evidence in favour of there having been an ocean, with shorelines forming topographical ledges (as in the figure on the previous page).

This is still an open question, however. Why, if Mars had a carbon dioxide atmosphere, did the ocean not absorb the gas to form carbonates (for example, chalk)? At least, no such rocks have been detected on Mars by planetary probes. Were the oceans, if any, too acid for such processes?

7 The mystery of Mars' past climate

A temperate climate, in spite of a young and fainter Sun

Although planetologists are still debating how ancient lakes and river valleys actually formed, they agree that a slow process of erosion by running water would have required fairly warm and wet conditions, if only episodically. That such a climate might have existed is surprising, for two reasons. On the one hand, Mars is quite a long way from the Sun; and on the other, as the experts tell us, the youthful Sun was, at the epoch in question, 25% less luminous than it is today. In such conditions the theoretical mean temperature of the planet would

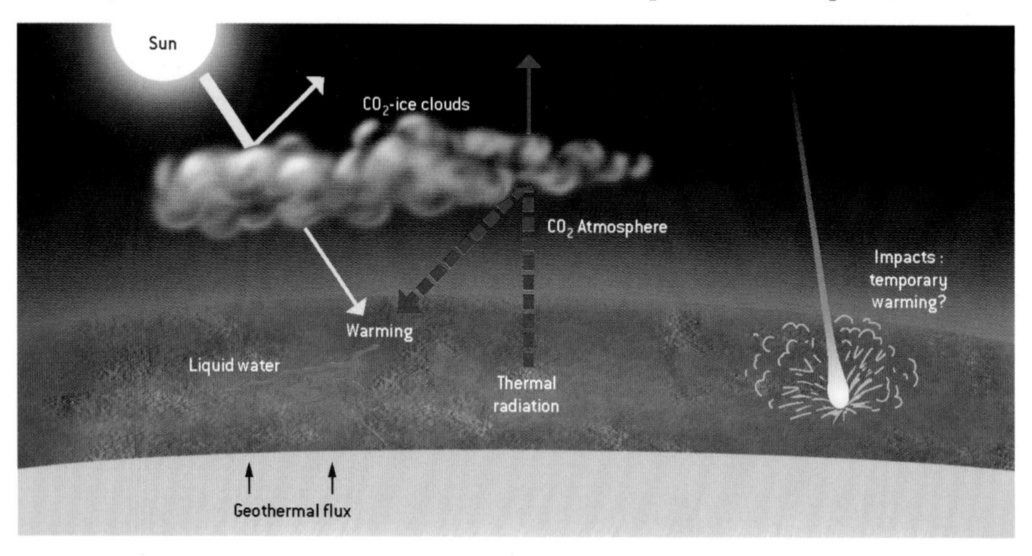

What would conditions have been like on Mars 3.5 to 4 billion years ago? The primitive Martian climate was warmed not only by the greenhouse effect due to carbon dioxide and water vapour (and other less chemically stable gases), but also by the presence of CO_2 ice clouds reflecting thermal radiation towards the surface. Acting like a reflecting 'survival blanket', such clouds would have significantly limited the cooling of the planet . Also, the recently formed (and therefore hot) interior of Mars created a geothermal heat flux 5-10 times more intense than is observed today. This contributed to the warming of the sub-surface and the maintenance of a hydrothermal system which probably played a key role in the establishment of the Martian climate. Finally, the cataclysmic impacts that occurred throughout Mars' early history may occasionally have led to warming of parts of the planet.

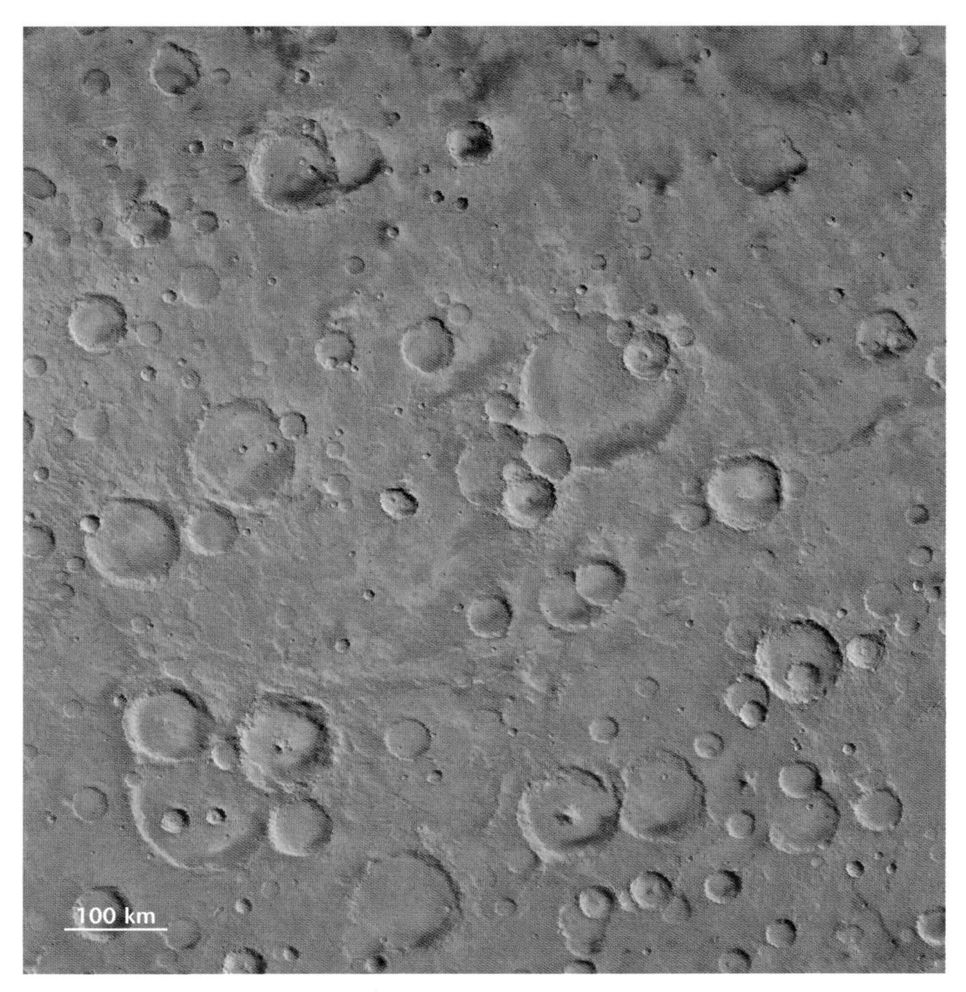

Erosion: slow today, much more intense in the past. A view of Noachis Terra, one of the most ancient terrains of Mars and exposed to erosion for the last 3.8 billion years. This image shows an area 900 kilometres across. As well as the presence of ancient 'fluvial' valleys, a careful study of the image reveals the fact that most of the small craters (less than 15 kilometres in diameter) have been erased, while larger craters have undergone substantial erosion. Craters found in regions with hardly more recent surfaces are much better preserved from the effects of erosion. Such observations show that, sometime in the distant past, erosion was thousands of times more intense. Mars doubtless experienced winds and precipitation generated by a dense atmosphere, which has now disappeared. (Courtesy NASA.)

have been around –75°C, in the absence of an atmosphere. How then can we explain the presence of a 'warm and wet' climate? The answer – a strong greenhouse effect.

A planet is warmed as it receives solar radiation; it is cooled when infrared radiation is emitted into space. The greenhouse effect occurs when cool gases

Former oasis Kasei Valles drains into the sea of Acidalia Planitia, in this re-creation of the Martian surface 3.8-3.5 billion years ago by Kees Veenenbos.

within the atmosphere allow incident solar radiation to pass through them, then absorb and trap the infrared (thermal) radiation emitted from the surface, thereby limiting planetary cooling.

The primitive Martian atmosphere must have been composed mostly of carbon dioxide, with a little nitrogen, argon and water vapour. At that time, there would have been enough CO_2 to maintain an atmospheric pressure of about 1 bar, similar to that on Earth. In such an atmosphere, only the CO_2 and the water vapour would be able to contribute to the greenhouse effect. It has, however, been shown that these two gases would not have been capable on their own of generating temperatures in excess of 0°C. To solve this problem, it has been necessary to consider the presence of other greenhouse gases such as ammonia, methane and sulphur dioxide. This mixture seems unlikely, since these gases would have been chemically unstable in a primitive Martian atmosphere, unless permanently produced by volcanic activity.

It seems more likely that Mars was warmed by clouds of carbon dioxide ice. We can envisage a dense atmosphere, rich in CO_2, with a tendency to condense into clouds of CO_2 ice. These clouds would have been able to reflect and effectively trap thermal radiation emitted from the surface, thereby considerably warming the atmosphere (see diagram on page 68). We might therefore picture a youthful Mars swathed in dense clouds, below which rivers flowed in perpetual semi-darkness. Far more evidence is needed to support such a scenario.

8 The theory of life on Mars

Life appeared on Earth 4 billion years ago. And on Mars...?

The oldest known sedimentary deposits on Earth date from 3.5 billion years ago, yet they already show evidence of biological activity. How could life have emerged so early in our planet's history? Nobody knows. We still cannot bridge the gap between non-living chemical compounds and the most basic life forms known. What is more, the rocks themselves offer no real record of such a transition. It is however the case that all theories require the presence of liquid

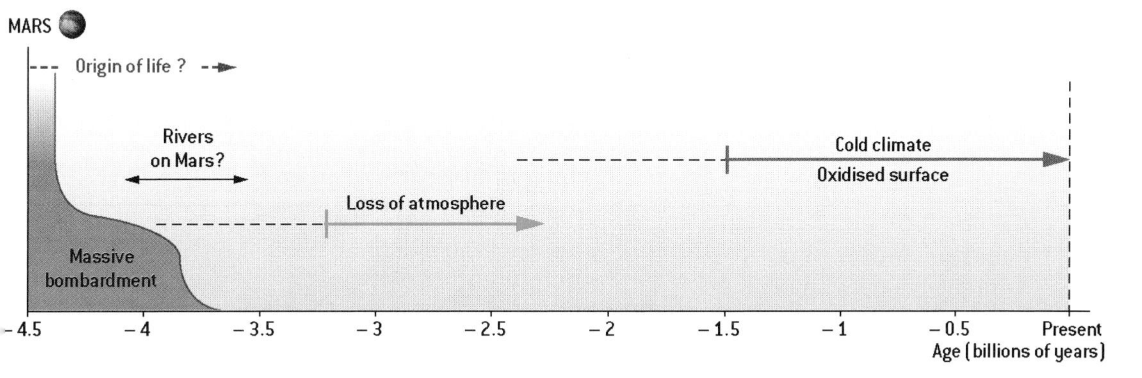

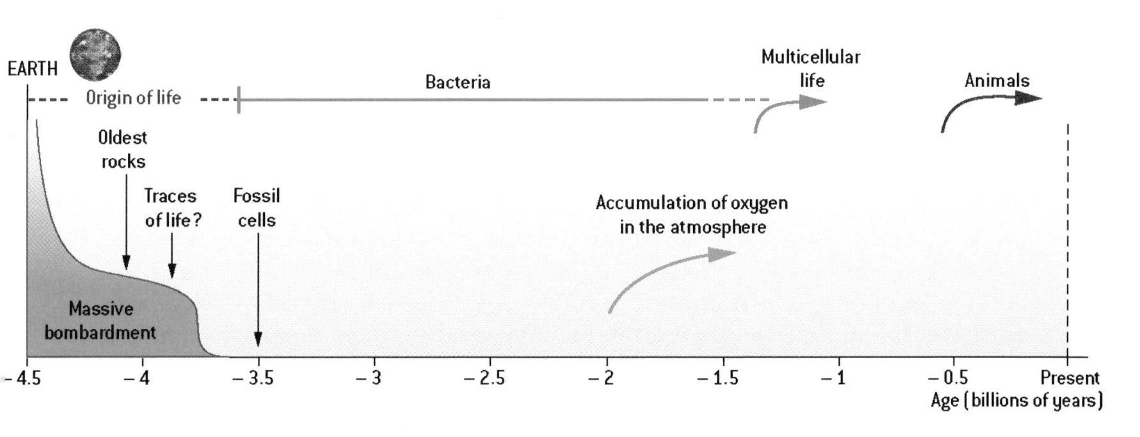

The evolution of life on Mars and on the Earth.

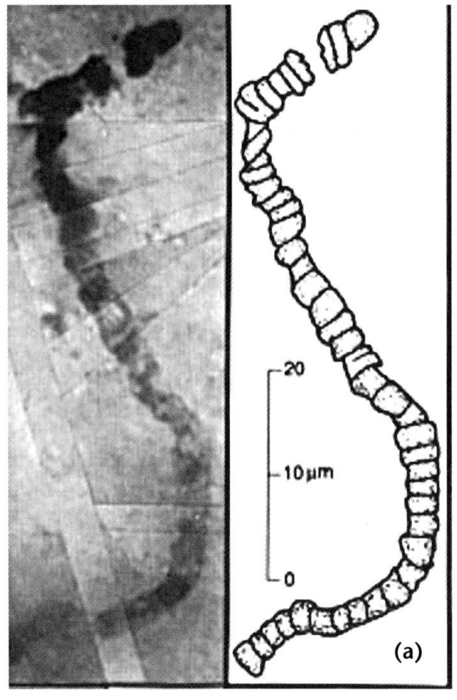

First traces of life on Earth. Evidence for life forms on Earth more than 3.5 billion years ago has been found in very rare sedimentary rocks which are still exposed. Fossils of bacteria (a) a few microns long (microfossils) have been identified and dated, though some controversy still surrounds the findings. Though debate still continues among specialists, stromatolites, structures of characteristically biological origin, as seen in image (b), confirm the presence of bacteria at that epoch. Stromatolites are produced by bacterial colonies of a primitive nature, which are rare today. (Images courtesy J. Williams Schopf, Univ. Calif. Los Angeles (a), and F. Bavendam, BIOS (b).)

water, and they agree on the likelihood that complex organic molecules were brought to this planet by meteorites and comets, impacting in vast numbers at that time. Volcanism and hydrothermal activity are also thought to have played a determining role, by providing energy and nutrients. It now seems that such conditions were present on Mars too. If life on Earth arose as a result of fundamental and repeatable physico-chemical processes, there is no *a priori* reason why life might not also have emerged on the red planet, 3.5 to 4 billion years ago.

What kind of life?

A study of Mars shows that conditions at its surface were such that liquid water could have been present there for a limited period only – at most, for a few hundred million years. If any life form evolved there, it could not have been anything but primitive. On Earth, the first multicellular organisms (and, *a fortiori*, the first animals and plants) appeared after more than two billion years of unicellular predecessors (see diagram on previous page). Was this because so much time was needed for the necessary genetic evolution, or was it because physical and chemical conditions on Earth (for example, amounts of oxygen) were incompatible with the existence of more complex creatures? Perhaps things did not occur in the same way on Mars, and some optimistic scientists have speculated that more evolved forms of life could have existed on Mars at the same epoch.

Did life on Earth come from Mars?

Mars, being smaller than the Earth, probably cooled much more rapidly. Also, it did not experience the same kind of tremendous impact which led to the formation of our Moon, and probably stripped the Earth of its early atmosphere, about 4.5 billion years ago. It could be, then, that life appeared on Mars before it appeared on Earth. In its early history, Mars suffered vast numbers of meteoritic impacts. These would often have ejected fragments of the planet's surface into space. Many of these fell on Earth (and may still do so). Inside these fragments, conditions may be such that various types of spores could survive. One fascinating and plausible hypothesis is that Martian organisms could have crossed interplanetary space and 'seeded' our planet 4 billion years ago.

9 Hunting for fossils

How can we search for traces of life that disappeared billions of years ago?

Exploring the Martian terrain in a few decades from now? (Illustration for NASA by Pat Rawlings, SAIC).

The discovery that life once developed on Mars would be one of the greatest scientific and philosophical events in human history. Merely demonstrating its absence would be of interest, as we ask ourselves the question: why, on two planets exhibiting similar conditions, did life emerge on one and not on the other? The fact remains that Mars presents a much better arena of research than Earth, where traces of almost everything that occurred more than three billion years ago have been erased. On the red planet, there may lie, awaiting discovery, a clue to the 'missing link' between non-living and living matter. But what should we be looking for?

Fossils

Normally, living cells decompose rapidly after death, and all traces of their morphology are lost. Nevertheless, it has often been possible to reconstitute the

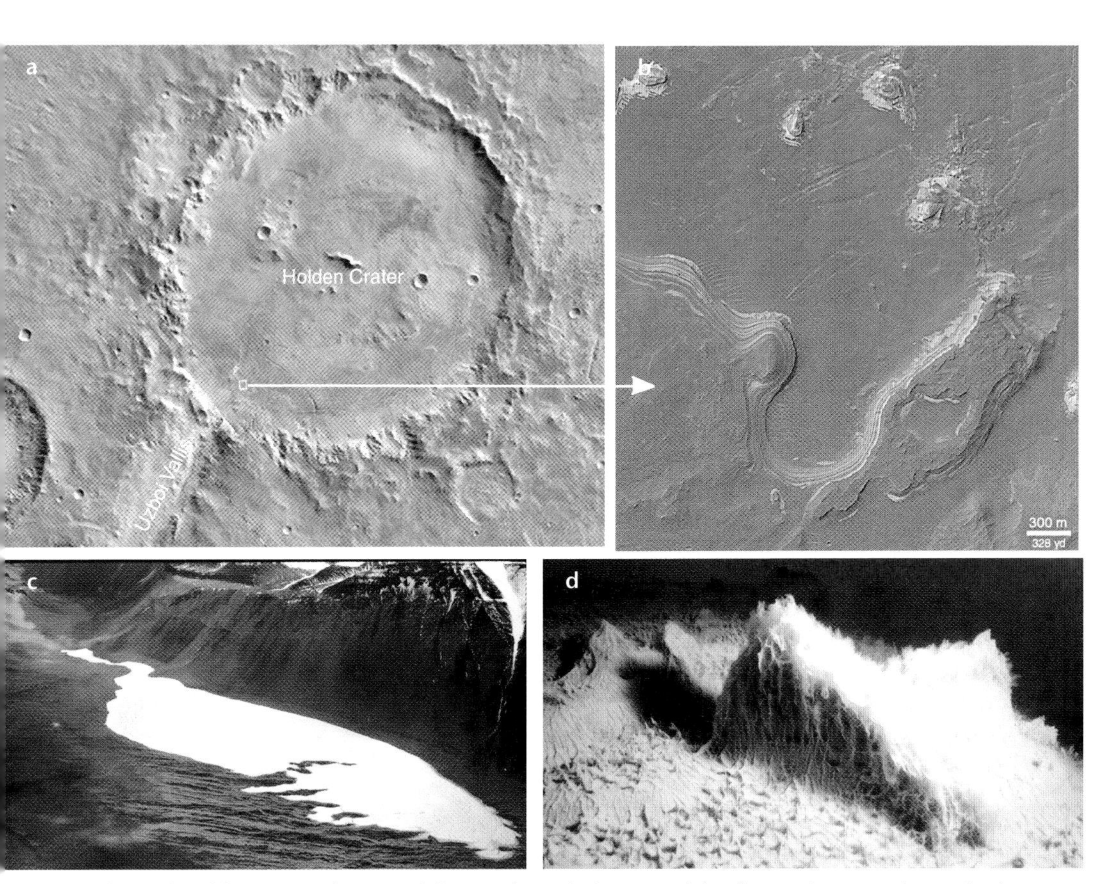

Where should we start the search? Now that missions capable of returning samples to Earth are being prepared (see pages 214–216), many scientists ask this question. Sedimentary basins, where water may have accumulated, are favoured sites. An example is Holden Crater (a, b), 1000 kilometres to the north of the Argyre Basin. (Courtesy NASA, JPL, Malin Space Science Systems.) Ancient lake sites are of particular interest, and it is possible that any Martian life there might have continued in a stable habitat after the average temperature of the site fell below 0°C, thanks to a thick covering of permanent ice. Such sites exist in the Antarctic dry valleys (c). They have been the object of many studies by NASA specialists. In spite of the absence of rain, and the sub-zero temperatures (except for a few days every year), these Antarctic lakes contain surprising numbers of ecological niches (d): bacterial colonies, several tens of centimetres across, carpet their depths. (Courtesy D. Andersen, SETI Institute.)

shapes, appearance and organisation of many ancient microbes, thanks to mineral fossilisation. What little is known of this process involves silicates in solution in water gradually being deposited on the organic molecules of cells, forming a solid mineral matrix, capable of conserving details as subtle as, for example, the texture of a bacterial membrane. It is quite possible for such a process to have occurred on Mars also. So, if any life form ever emerged there, we may one day be able to find clues to its morphology.

Organic material

Another avenue of investigation is the search for the remains of organic matter. Unfortunately, the Viking landers showed that the soil of Mars is chemically very 'aggressive', containing oxidising compounds which have probably destroyed any organic molecules in the topmost few metres of its surface (see pages 83–85). Should any such molecules be found at greater depths, the thorny problem will then arise of determining whether their origin is mineral or biological. Now, experiments carried out on Earth show that most complex molecules characteristic of living matter ought to have totally disappeared on Mars after 3.5 billion years. There remains however the possibility that the permanently frozen soil might be capable of preserving certain particularly hardy compounds.

Chemical anomalies

Any life that once existed on Mars might have been able to concentrate certain chemical compounds in quantities difficult to explain as the result of non-biological processes. On Earth, pyrite, magnetite and phosphates are found in strong concentrations within living cells. They are used as 'markers' of biological activity. Also, all living things on Earth show a 'preference' for carbon-12 rather than carbon-13, which is a chemically identical (but heavier) isotope. This phenomenon could be detected on Mars, just as it is on Earth.

10 A unique witness, 4 billion years old

Meteorite ALH84001: a witness to the youth of Mars – and to its early life forms?

Almost all of the SNC meteorites from Mars (see pages 30–32) are relatively recent basaltic rocks, aged between 180 million and 1400 million years. Only one of them is different: ALH84001, discovered in the Allan Hills (hence the abbreviation) in Antarctica, in 1984. This rock was most probably formed more

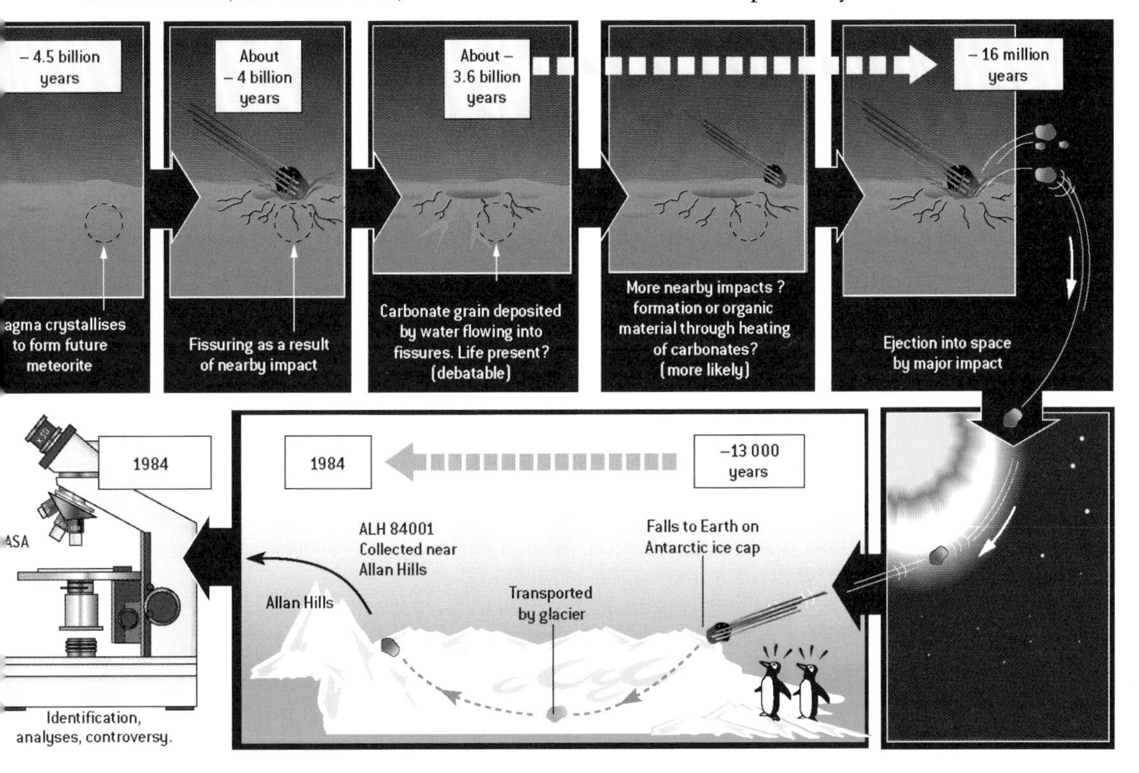

The amazing story of a Martian rock. The history of this meteorite has been pieced together using isotopic dating methods on the elements fixed in the rock by crystallisation (values are 4.5 billion years ago for volcanic rock, and 3.6 billion years ago for carbonates), heating, following impacts (4 billion years), and the exposure of the rock to cosmic rays during its journey through space. Analysis indicates that cosmic rays began to create new isotopes 16 million years ago, and that the process ceased abruptly 13 000 years ago.

Analysis of meteorite ALH 84001. **(a)** When it was found, this object was covered with a fusion crust characteristic of meteorites which have been heated by friction during their passage through the Earth's atmosphere. **(b)** Cut in two, ALH 84001 displayed numerous fractures and voids. **(c)** Through the microscope, strange spherical carbonate inclusions, of diameters 100-200 micrometres (0.1-0.2 mm) were observed. These bodies constitute less than 0.5 percent of the whole. **(d)** Scanning electron microscope image of the surface of a carbonate: tiny stick-like objects composed of a chain of supposed 'fossils' measuring less than 0.1 micrometre. (Images courtesy NASA (a), NASA, JSC (b), (d), and John W. Valley, Univ. of Wisconsin-Madison (c).)

than 4 billion years ago in a temperate and chemically reducing environment, very different from the oxidising one found on Mars today. ALH84001 shows fissures through which water seems to have flowed about 3.6 billion years ago, depositing, most noticeably, carbonate nodules. These are about a tenth of a millimetre in diameter. There are also organic molecules within this meteorite. It is therefore a precious and unique witness of a period when Mars was an oasis.

After its identification in 1993, this exceptional meteorite became the object of world attention. In 1996, a team of American researchers under David McKay of NASA made international headlines by suggesting that ALH84001 might contain traces of extraterrestrial microbial life. But, where was the evidence?

Firstly, there was the presence of complex organic compounds (polycyclic aromatic hydrocarbons, or PAHs), which could possibly have been the remains of the decomposition of microbes. Secondly, the appearance and chemical nature of the carbonate nodules was reminiscent of deposits left by certain earthly bacteria, showing in particular rich layers of iron oxides such as magnetite. And lastly, 'fossils', less than one micrometre long and bearing a striking resemblance to bacteria, could be seen under the electron microscope.

Years went by, and many scientists called McKay's results into question. Could the 'microfossils' be products of abrasion and alteration of the surface during the process of observation? And might not the PAHs, carbonate nodules and magnetite deposits be due to natural mineral processes? In the end, although it had been proved that the carbonate deposits were indeed ancient, and of extraterrestrial origin, it was thought probable that at least some of the organic matter resulted from contamination of the meteorite during its 13,000-year sojourn in the ice of Antarctica. Even if ALH84001 is not formal proof that life once existed on Mars, it confirms that the planet, in its early years, offered clement and probably moist conditions, in which any organic compounds would not have been immediately destroyed through oxidisation.

11 The end of an oasis: Mars loses its atmosphere

Why did Mars not follow the same path as Earth?

Mars seems once to have possessed a mild and humid, Earthlike climate. Now, it is a cold and dry planet, with a tenuous atmosphere ruling out the presence of rivers. Many researchers agree that Mars reached the crucial point in its destiny when it lost most of its atmosphere. How did this happen?

Meteoritic impacts

For the first billion years of their existence, Mars and the Earth were in collision with asteroids kilometres across, and the violent impacts could have ejected much of Mars' early atmosphere into space. The Earth, with its greater gravity, was better able to retain its gases. Even so, the 'loss-by-impact' theory cannot totally explain why Mars today has such a thin atmosphere. Calculations based on what we know of atmospheric erosion suggest that, at the end of the period of intense bombardment, the Martian atmosphere was much more massive than it is at present.

Reaction with the surface

In the presence of liquid water, carbon dioxide in the atmosphere reacts chemically with surface rocks to form carbonates, thereby reducing the volume

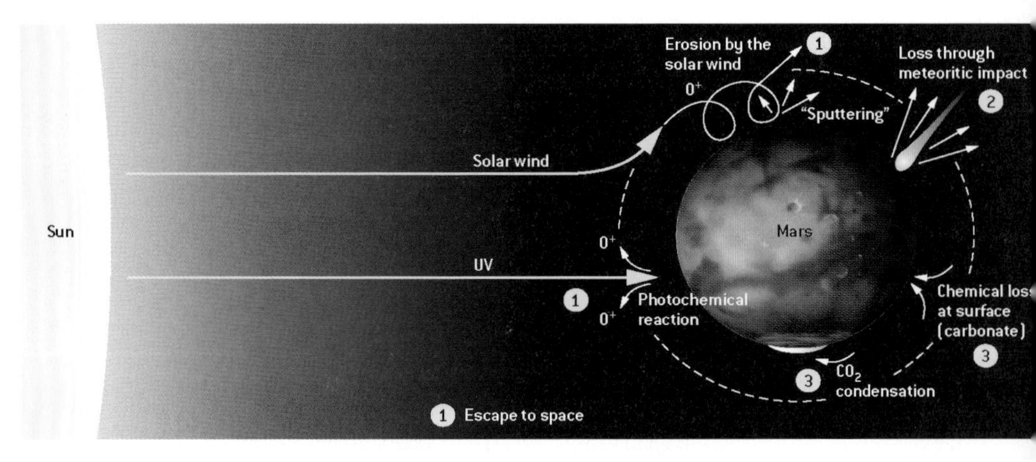

Diagram of the three processes leading to the loss of the Martian atmosphere. (1) Escape into space; (2) meteoritic impacts; (3) physico-chemical reactions at the surface.

Different destinies. Mars (right) is now only a vast desert, having failed to retain its atmosphere. This outcome is explained not only by Mars' distance from the Sun, but also by internal factors: the planet's low gravity, and the lack of both active plate-tectonics and a magnetic field. In brief, it is just too small, compared with Earth (left). (Courtesy NASA, JPL.)

of the atmosphere. On Earth, this process is balanced out by the constant 'recycling' of rocks as tectonic processes proceed. Carbonates drawn into the depths of the Earth give off carbon dioxide, which is later returned to the atmosphere by volcanoes. On Mars, where plate-tectonic activity is absent, most of the atmosphere would probably have been transformed into carbonates at the surface. The question then arises: where are these carbonates? No definite trace of them has yet been found, even by the OMEGA mapping spectrometer aboard Mars Express, which should be able to detect just small amounts of them. Only a few traces were found within the ALH84001 Martian meteorite (see pages 77–79). Has the Martian climate destroyed all vestiges of carbonates on the planet's surface?

It is possible to envisage a scenario where carbonates never even formed. If Mars' primitive environment encouraged the deposition of sulphates (see pages 56–61), surface water would have been relatively acidic, and most kinds of carbonates would have been unable to form.

Escape to space

A planet must have sufficient gravity to be able to retain its atmosphere: on a relatively small body such as the Moon, for example, the simple thermal motion of a gas molecule is enough to ensure its escape to space. On Earth, however, this kind of loss is negligible, except in the case of the lightest gases, hydrogen and

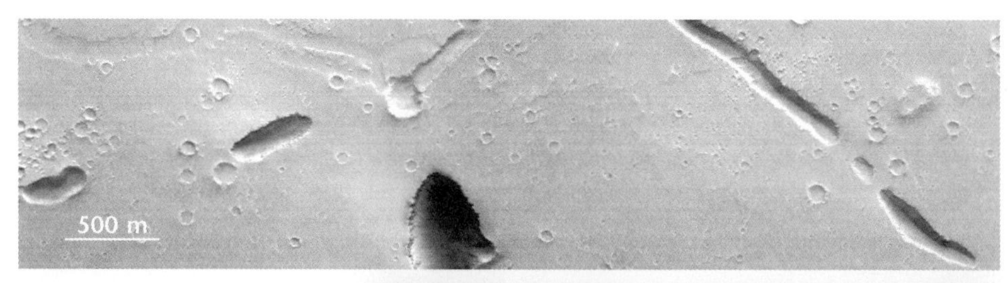

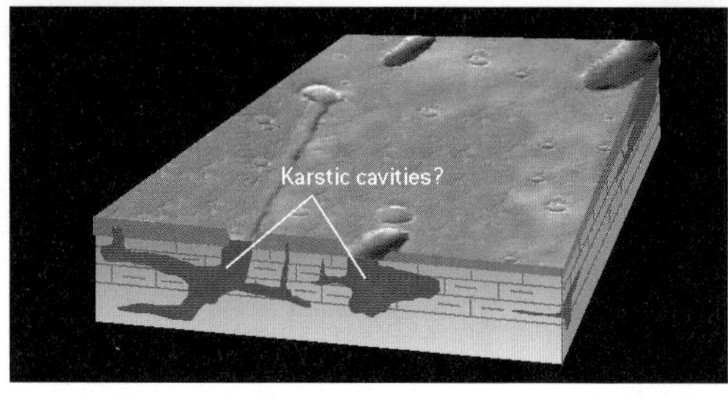

Hidden carbonates on Mars? In theory, part of the ancient Martian atmosphere might have been locked into the Martian sub-surface in the form of carbonates. These carbonates are absent from the surface itself. However, there are some examples of ancient 'rivers' (e.g. here, Hebrus Vallis) which have apparently carved their beds both at the surface, and in the sub-surface by dissolving rocks to form a 'karstic cavity'. This is a morphology found in limestone areas, rich in carbonates, on Earth. Does the sub-surface material in this region consist of a layer of carbonates now buried beneath dust? (Courtesy NASA, JPL, Malin Space Science Systems.)

helium. What is the situation on Mars? The red planet is massive enough to ensure that carbon dioxide and water vapour are not lost in large quantities. There are other processes, though, which lead to the ejection of atoms and molecules. Reactions resulting from solar ultraviolet radiation, and the interaction of the solar wind with ions high in the atmosphere, are both contributory factors. In the case of the Earth, its magnetic field deflects the solar wind above the level of the atmosphere. Mars has been without a magnetic field for aeons, and oxygen ions, strongly accelerated in its upper atmosphere by the magnetic field of the solar wind, can encounter CO_2 molecules and eject them into space.

12 Life on Mars today

Could primitive life forms have survived on such a hostile world?

The Viking probes' search for life. The detection of life on the Martian surface was one of the main objectives of the Viking probes (1976). An articulated arm, capable of digging into the surface, gathered samples of Martian soil. The samples were subjected to three investigations: the Pyrolitic Release, Gas Exchange and Labelled Release experiments (see diagrams on page 84). (Courtesy NASA.)

Mars today is cold and dry. Without a protective ozone layer above it (see pages 125–127), its surface is exposed to intense ultraviolet radiation, which would prove lethal for most living organisms (at least, for organisms such as those found on Earth). Also, the upper few metres of the soil seem to have been sterilised by highly oxidising compounds present in the atmosphere, the equivalent of the oxygenated water used to kill microbes. In such conditions, assuming that life ever did get started on Mars, how could it have survived until today? Even the most optimistic specialists are somewhat guarded on this subject. Remember, though, that exploration of our own planet has revealed to

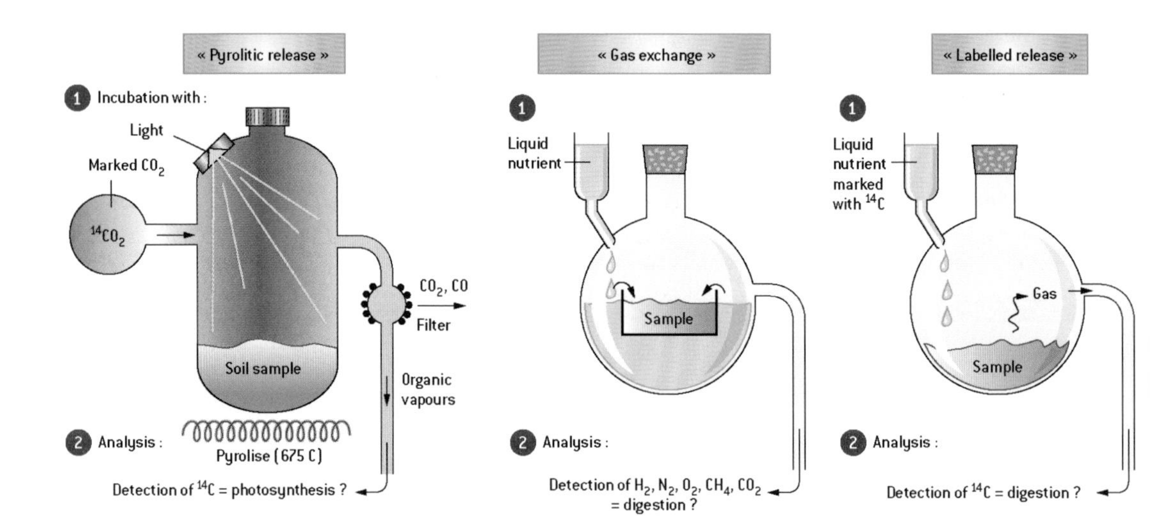

Analysis of Martian soil by the Viking probes. The **Pyrolitic Release** experiment was designed to check if possible Martian organisms would assimilate carbon dioxide, as certain terrestrial organisms do (photosynthesis). No such activity was detected. The **Gas Exchange** experiment was designed to detect gas produced by micro-organisms having digested organic matter. Quite a lot of oxygen was detected, but as it was still being produced after sterilisation by heating the sample to 145°C, it was concluded that the oxygen came from the chemical oxidisation of the material. Rather than detecting life, the experiment showed that the surface of Mars is an oxidising and sterilising milieu, rather like some medical disinfectant. The **Labelled Release** experiment was designed to detect, by radioactive marking, the transformation of organic matter into carbon dioxide. The result proved positive: there was indeed a transformation, which ceased as soon as the sample was heated – exactly what one might expect if bacteria were present. However, the verdict was finally negative, because another instrument, a gas chromatograph coupled with a mass spectrometer, capable of detecting organic molecules at a concentration below one part per billion, found nothing, to the surprise of all concerned.

In the light of the other experiments, most of the experts, though not all, concluded that the labelled release experiment was evidence of the chemistry of inorganic processes in an oxidising environment. The surface of Mars seems to have been completely purged of all natural organic matter by ultraviolet radiation and oxidisation.

us that some micro-organisms can tolerate the most extreme environments. Though subjected to very high pressures, to temperatures above 100°C, and to very acidic or alkaline surroundings, or subsisting on hydrogen sulphide or iron, still they survive. In 1995, several kilometres below the Earth's surface in Oregon, organisms with an iron-based metabolism were discovered living in basaltic rock.

There is one thing, as far as we know, which seems indispensable to the survival of all life forms: liquid water. If life has survived until the present day on Mars, we will have to look for it in those niches where liquid water is still present. One possibility is several kilometres underground, where the planet's internal heat source ensures that temperatures remain above 0°C. 'Interstitial' water could harbour life, just as has happened in the Oregon example. Nearer to the surface, Martian volcanism, which, though on the decline, still occurs, probably creates warmer places. Just as on Earth, organisms might exploit the relatively rich chemical resources found in such locations. It is possible to envisage bodies of liquid water having formed below the permanent Martian polar caps, in the manner of the subglacial Lake Vostok in Antarctica.

It must be stated, however, that for life to have held out this long on the red planet, Martian organisms would have needed a stable environment in their 'bio-niches'. This seems improbable, given a planetary evolution of billions of years: but what new insights will the next discovery bring?

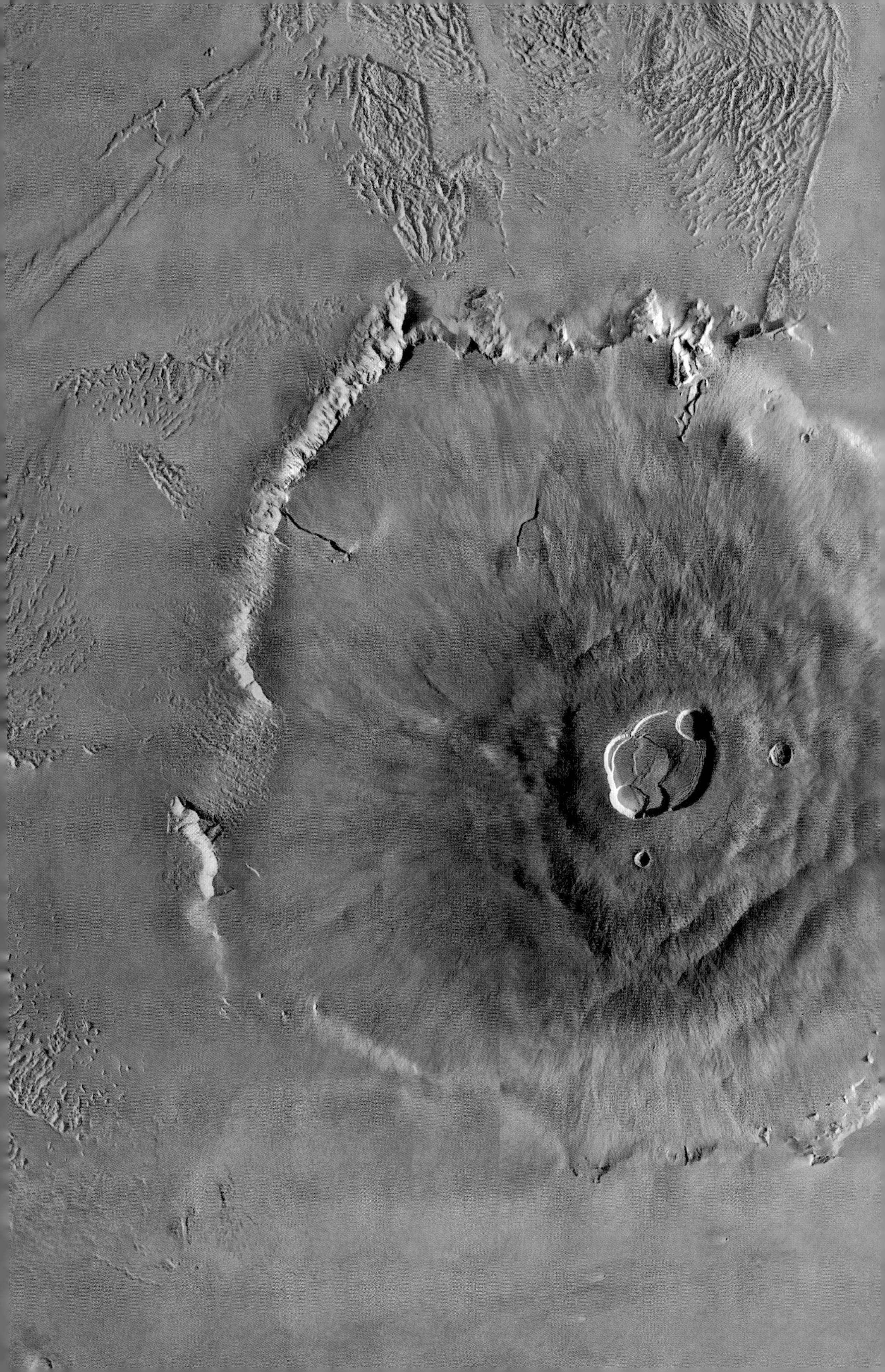

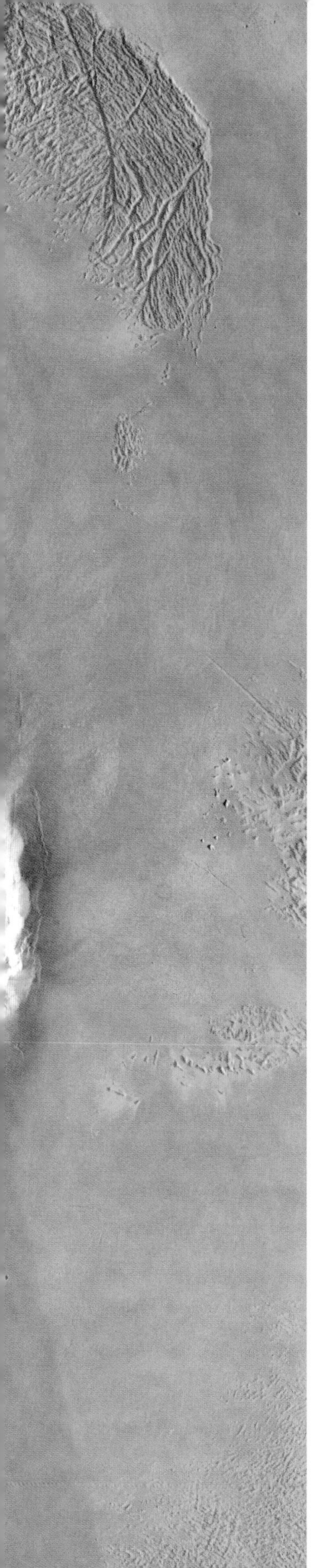

The slow metamorphosis

−3.8 to −0.1 billion years

Now, a journey through time on Mars, across billions of years: as centuries passed, lava, ice and sometimes running water worked together to fashion the surface, creating the Mars that we see today. Only a few glimpses of that history are available to us, worked over by geologists in the light of new discoveries. Each space mission unfailingly brings us a stack of surprises – and challenging questions.

A giant volcano on Mars: Olympus Mons rises 23 kilometres above the surrounding plains (Viking image). (Courtesy USGS.)

1 A sub-surface motor: convection in the mantle

All geophysical activity on Mars is linked to the loss of its heat

In its infancy, Mars displayed its energy: its giant volcanoes testify to the fact that Mars was once hot enough for massive volcanism to occur. Much of its early and even present atmosphere and water could have come from the mantle in the form of volcanic gases. However, this vigorous eruptive phase soon gave way to a process whereby internal heat was gradually lost through convection.

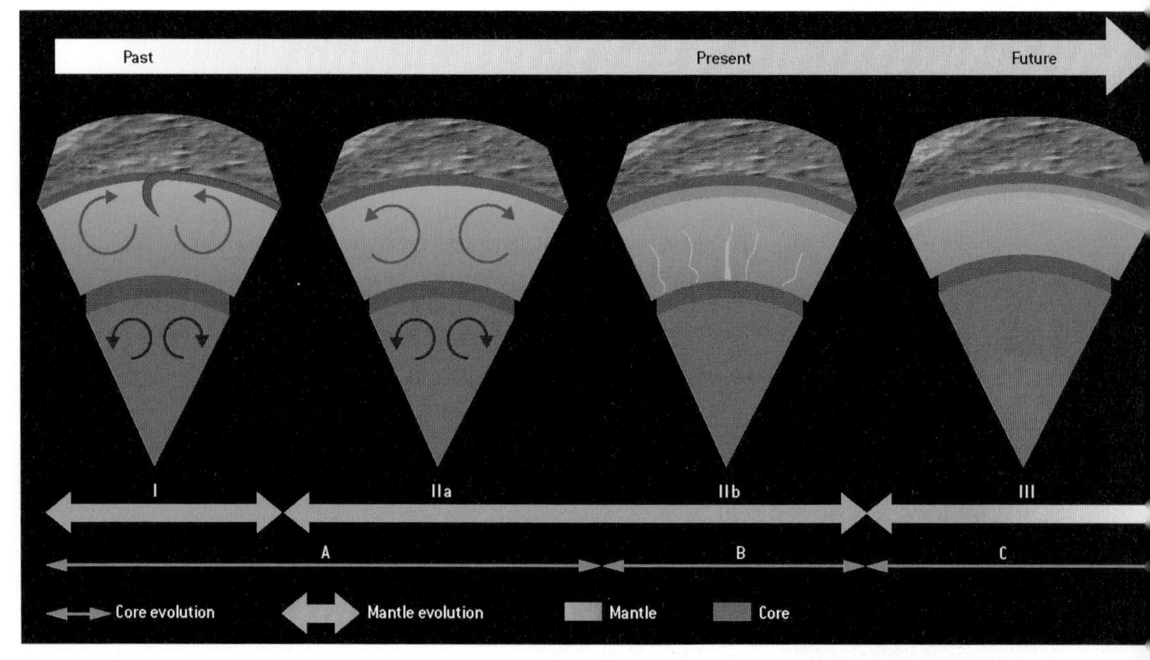

The history of Mars' mantle and magnetic field. Between 4.4 and about 4.0 billion years ago, strong convection currents were occurring in Mars' mantle (see diagram: mantle epoch I), causing renewal of the surface (a hypothesis as yet unconfirmed). Loss of heat by convection from the mantle caused convection within the iron core, with movements creating a magnetic field (core epoch A). After a few hundred million years, the convection decreased and the crust became stable at the surface. The planet moved into its 'single-plate' tectonic phase (epoch IIa). The convection became less energetic and only a few 'hot spots' remained (epoch IIb). Within the core, convection gave way to conduction as the agent of heat loss, and cooling proceeded. Movement within the core, and the induced magnetic field, disappeared (epoch B). Cooling will continue in the future, and convection will end in the mantle (epoch III), with solidification of part of the core (epoch C). Mars will then be tectonically dead.

The origin of convection

Mars' internal heat originally came about as a result of accretion followed by the formation of an iron core, and was later maintained by radioactivity in the uranium, thorium and potassium present in the mantle (see Part 1). However, the amount of energy within a planet is not inexhaustible, and when all the energy due to the process of formation and to radioactivity has been converted into heat, the planet, in the cold environment of space, will cool. Convective movements within the mantle bring its internal heat from the core towards the

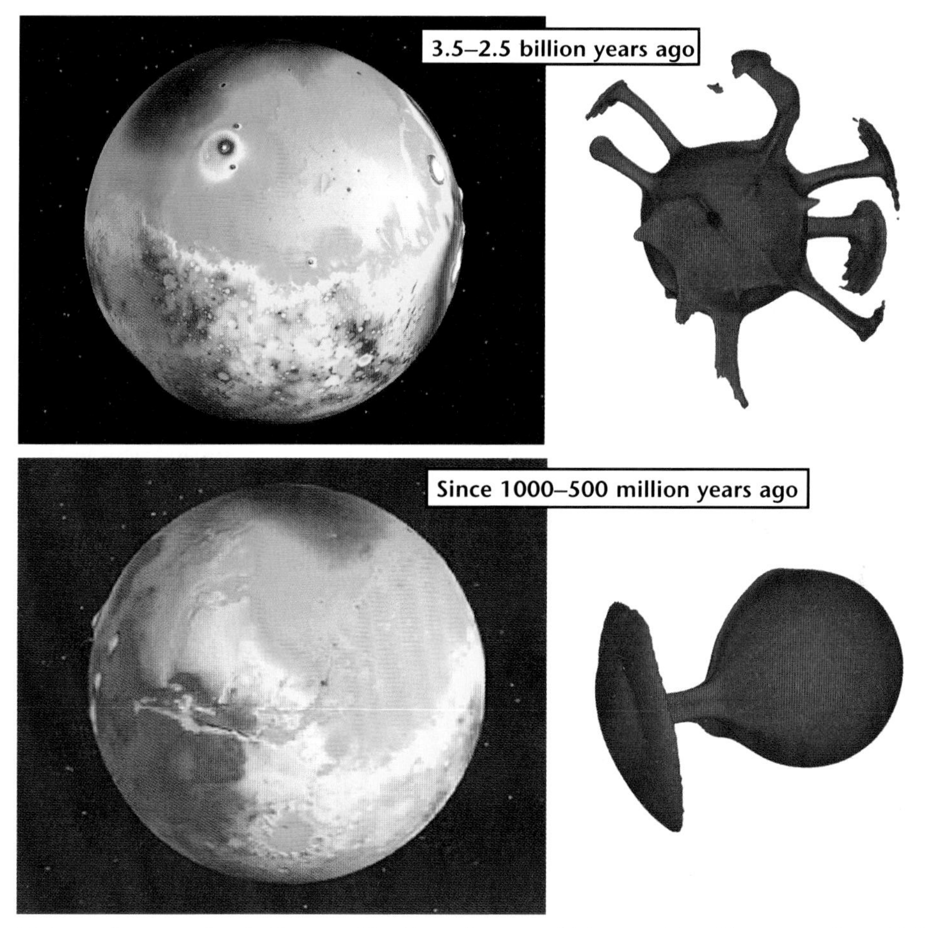

History of convection in the mantle. The volcanoes of Mars explained by convective activity within the mantle, based on a simulation of the mantle (right) by a team from Göttingen University. The hottest rocks are shown in red. Currently, as the Martian mantle is still heated, the lava 'hot spot' beneath Tharsis ought still to be active. This is what images of lava flows with almost no craters (and therefore very recent) on them seem to indicate (see pages 94–96). Is this lava plume still able to rise through the lithosphere? Only precise dating of the lava will provide the answer to this question. (Courtesy H. Harder and U. Christensen, Univ. of Göttingen.)

surface. In the same way that water moves when heated in a saucepan, the solid Martian mantle deforms and contorts, slowly bringing warmer material from the depths into contact with the cold layer near the surface (though in reality convection is confined between the core and a few hundred kilometres below the surface). Convection within the mantle cools the surface of the core, and triggers further convection of the core. The 'chain' of cooling is then established, and will shape the face of the planet throughout its history.

The evolution of convection

For the first two billion years of Mars' existence, convection within the mantle proceeded very vigorously. Progressively, the most radioactive elements were purged from the mantle, either though disintegration or because they had risen into the crust in lavas. But convection did not cease altogether. The slow degradation of convection on Mars, which some theoretical models have stumbled over, is due to the crust, now containing radioactive elements, acting like a warming blanket atop the mantle. On Earth, between 30 and 40 percent of heat production has been concentrated within the continental crust. Now, Mars' ever-thickening 'carapace' has imprisoned its mantle inside a rigid and insulating layer of material, known as the lithosphere. Its depth varies between 500 kilometres in the south and 150-250 kilometres in the north. Mars has followed the path taken by the Moon, which has a lithosphere 800 kilometres thick and maintains its liquid core even to this day. The red planet too probably has a very hot mantle and a liquid heart, and some scientists have even proposed that the mantle, owing to the crust blanketing effect, might be partially melted beneath a few volcanoes, where young lava flows have been found by Mars Express.

2 Giant volcanoes on Mars

Volcanoes vie with each other to be the biggest

Among the most characteristic structures on Mars are its great volcanoes. They are many and varied, and some of them are enormous. They tell of recent geological activity. The planet's most imposing volcanoes are concentrated within the areas of Tharsis and Elysium. In the past, they have given out vast quantities of gases, including CO_2, SO_2 and water vapour, which probably led to climatic warming and sulphate deposition (see pages 59–61).

Shield volcanoes

The Tharsis region consists of a large upraised bulge, 5000 kilometres across and 5 kilometres high, upon which sit immense extinct volcanoes: Arsia Mons, Pavonis Mons, Ascraeus Mons and Olympus Mons. All these structures are similar in morphology to their terrestrial cousins, the shield volcanoes of the Hawaiian Islands. Their slopes are shallow (not exceeding 6° in inclination), and they have collapsed zones (calderas) at their summits. These structures are the products of the accumulation of very fluid lava flows, which explains their gentle slopes and the widths of their bases.

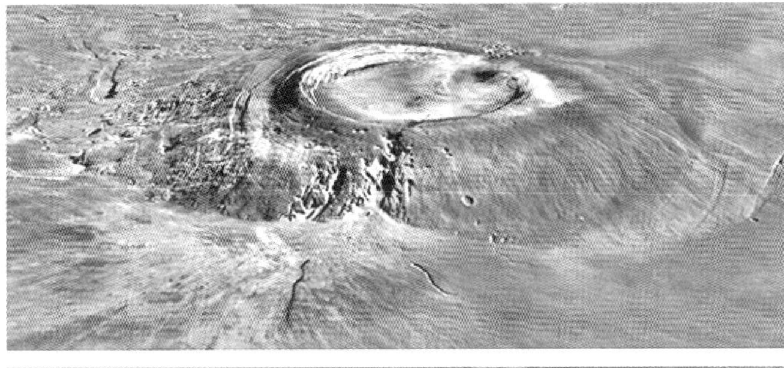

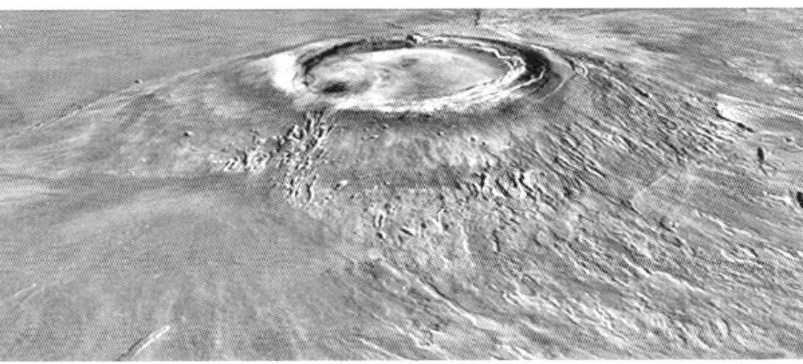

Shield volcanoes in the Tharsis region. The volcanic edifice of Arsia Mons, seen from two different directions. The vertical scale is exaggerated ten times. The diameter is 350 kilometres. (Courtesy NASA, MOLA Science Team.)

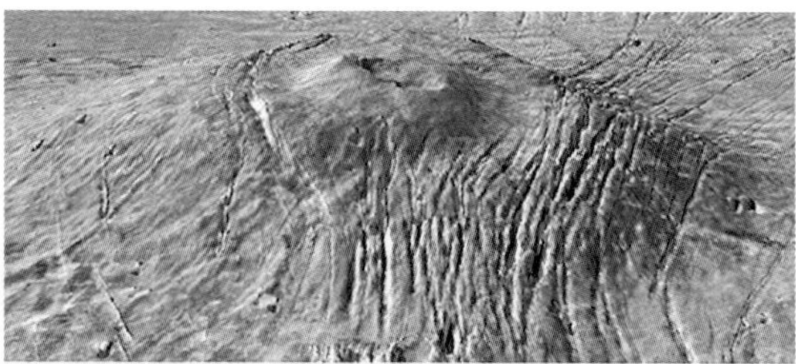

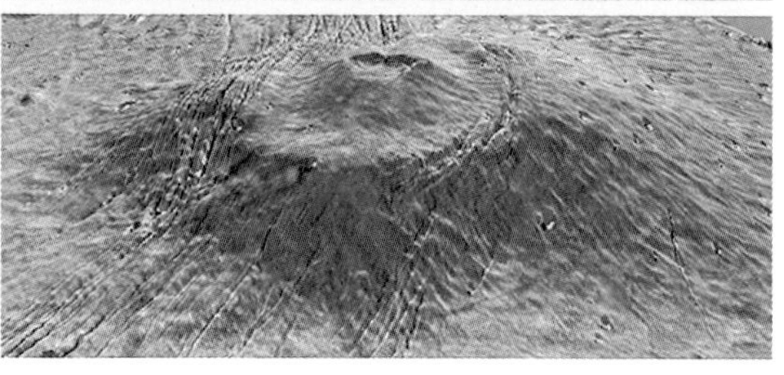

The Paterae. The image at left shows Apollinaris Patera, with its 80-kilometre caldera. (Courtesy NASA, JPL, Malin Space Science Systems.) A water-ice cloud has formed over the summit. The outflows of Apollinaris Patera can easily be imagined from the image: lava has partly covered and erased the escarpment at the base of the volcano. Above right, two 3-D views of Alba Patera. The vertical scale is exaggerated ten times. Alba Patera is a huge structure 1500 kilometres in diameter to the north of the Tharsis dome. (Courtesy NASA, MOLA Science Team.)

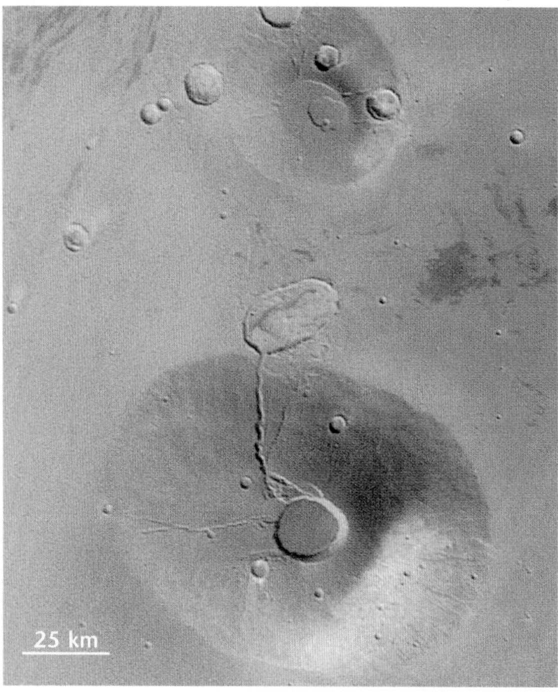

The Tholi. The image shows volcanoes Ceraunius Tholus (below) and Uranius Tholus (top). (Courtesy NASA, JPL, Malin Space systems.)

The Paterae

These pie-shaped volcanic structures may be hundreds of kilometres wide. They are older than the shield volcanoes of Tharsis. They are found near Tharsis (e.g. Apollinaris Patera) and in Elysium (Hadriaca Patera, Tyrrhena Patera and Amphitrites Patera). They have a central caldera, and lava flow channels on their flanks. Alba Patera (to the north of Tharsis) and Apollinaris Patera appear to have had explosive episodes of the hydrothermal type, triggered by encounters between their magma chambers and underground water. The resulting explosions were characterised by the ejection of 'pyroclastic' debris, and the thick clouds of ash known as *nuées ardentes*. It is thought that these ash falls, lying in vast, loose deposits called ignimbrites, were later dissected, as valleys sometimes many kilometres long were cut into them by erosion around the caldera. An 'effusive' (fluid lava) phase would have succeeded the explosive episodes.

The Tholi

These volcanoes are of a different kind again. They are situated mainly in the Tharsis region, near the shield volcanoes. They are smaller than the latter. Their steeper slopes (about 8°) may be the result of a more viscous kind of lava, and it is not out of the question that some of them exhibited a late, explosive phase.

3 Olympus Mons: the solar system's biggest volcano

A giant more than 20 000 metres high and 600 kilometres across

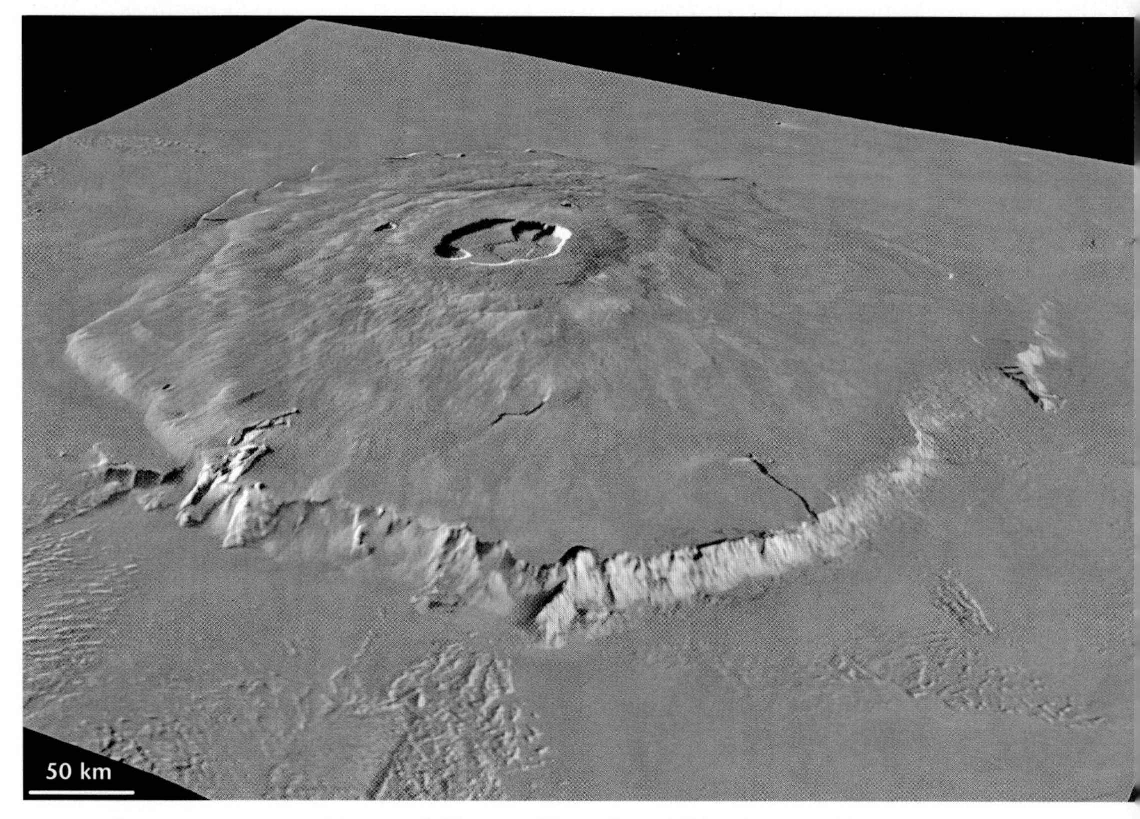

Computer-generated image of Olympus Mons, from Viking images. (Courtesy NASA.)

Olympus Mons is a colossus. Its volume is 50 to 100 times greater than that of the largest volcanoes on Earth. How was it formed? It is merely (!) an accumulation of lava flows from an underground reservoir of magma. On Earth, volcanoes of this type are known as 'Hawaiian', after those on the Hawaiian Islands, soaring to more than 7000 metres above the ocean floor. The magma chamber of giant Olympus must be at least three times bigger than that of a Hawaiian volcano; but the reason behind its enormous size is not to be found in the manner of its

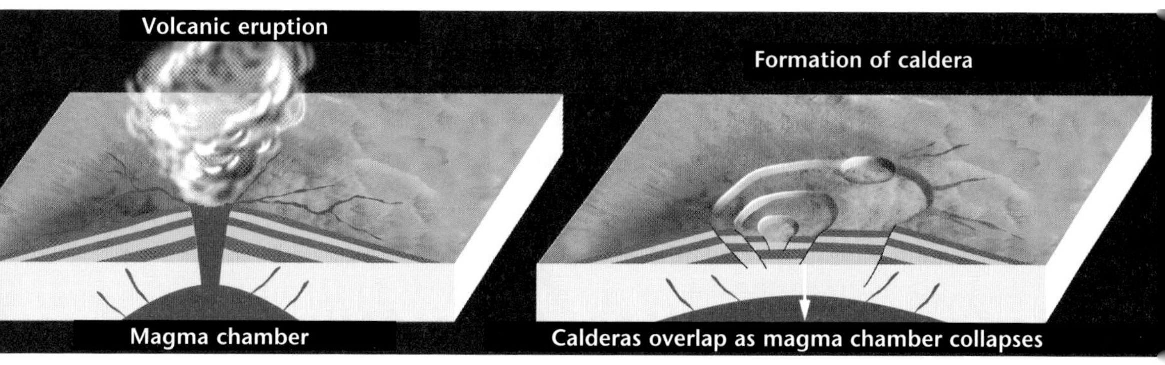

How does the shield volcano Olympus Mons work? Left: a deep magma chamber provides large volumes of fluid lava, which flow down shallow slopes. Right: The calderas are formed by the collapse of the magma chamber. Numerous circular faults are seen within them.

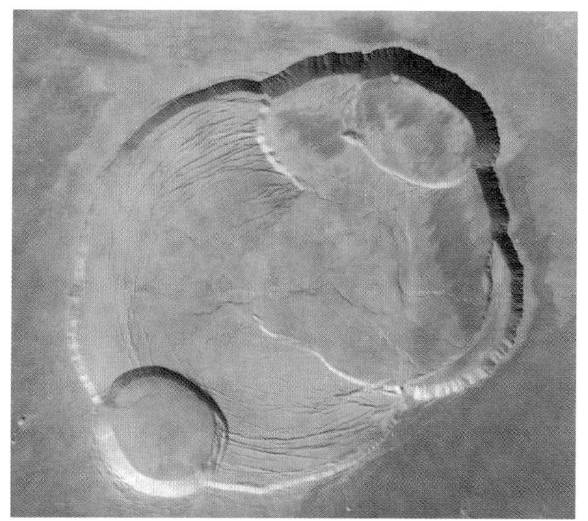

The calderas of Olympus Mons, 80 km across, photographed by Mars Express. They were formed during successive episodes between 200 and 100 million years ago. (Courtesy ESA, DLR, FU Berlin (G. Neukum).)

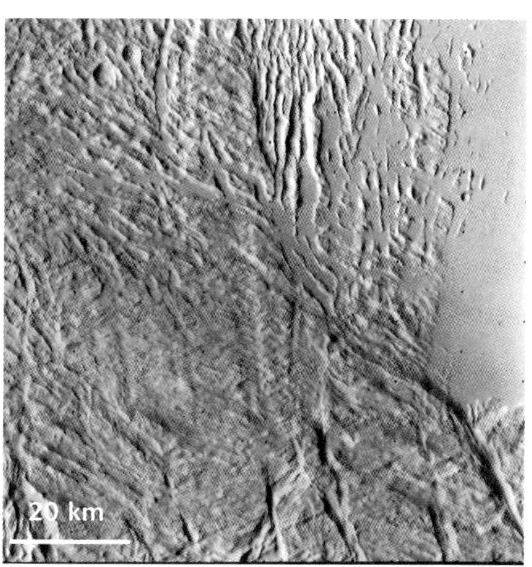

To the north-west of Olympus Mons, vast 'aureole' deposits extend for distances of the order of 700 kilometres (Viking image). (Courtesy NASA.)

construction, but rather in the length of time during which it occurred. Plate-tectonic movements on Earth displace the crust above the magma chamber as the lava flows; so, instead of one large volcano, a chain of smaller ones (as in the Hawaiian Islands) appears. Since tectonic activity is absent on Mars, layer upon layer of lava has emerged from Olympus Mons' magma chamber, accumulating over hundreds of millions of years.

Olympus Mons may resemble a Hawaiian-type volcano, with its 6° slopes, but it also shows characteristics associated with Icelandic volcanoes. For example, around its base is an escarpment 6000 metres high. This feature has probably been created by the action of the very weight of the volcano deforming the layers of lava and ash.

The caldera of Olympus Mons is 80 kilometres across and 2600 metres deep. It consists in fact of six overlapping calderas. The great mass of the volcanic dome causes slumping, which creates obvious cracks (see photo on page 95). To the north-west of Olympus Mons, vast aureoles stretch away to distances of as much as 700 kilometres. The origin of these aureoles is not well understood. They could be produced by erosion of the volcanic deposits, or intense fracturing. Other explanations invoke the existence of ancient glaciers (see pages 170–172).

The HRSC camera on the European Mars Express probe has recently refined estimates of the dates when the last volcanic episodes occurred on Olympus Mons. Crater counts suggest that the summit caldera was created between 100 million and 200 million years ago. The most recent lava flows on the flanks of the volcano are only about 2 million years old. So, Olympus Mons may possibly still be active.

4 Lava flows and lava plains

Lava has fashioned great expanses of Mars

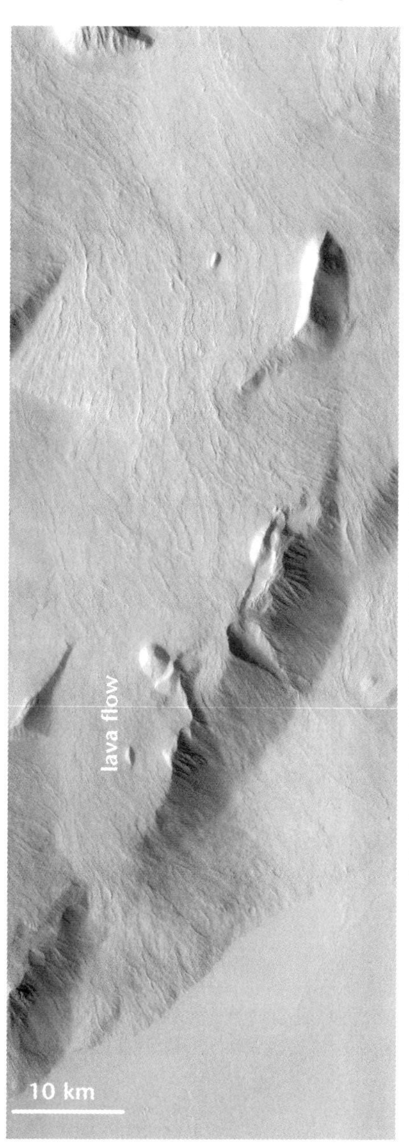

Volcanism on Mars is not limited to the building of mighty structures such as Olympus Mons. The fluidity of lavas varies to create many other features.

Rivers of lava

In the areas (Tharsis, Elysium, Hellas Planitia and Syrtis Major) surrounding most Martian volcanoes, there is evidence of numerous lava flows, some of them recent, which have travelled for hundreds of kilometres from the flanks of these volcanoes. These flows were of the same nature as those of the shield volcanoes of Hawaii, but in volume they may have been five times larger. Here again, the absence of

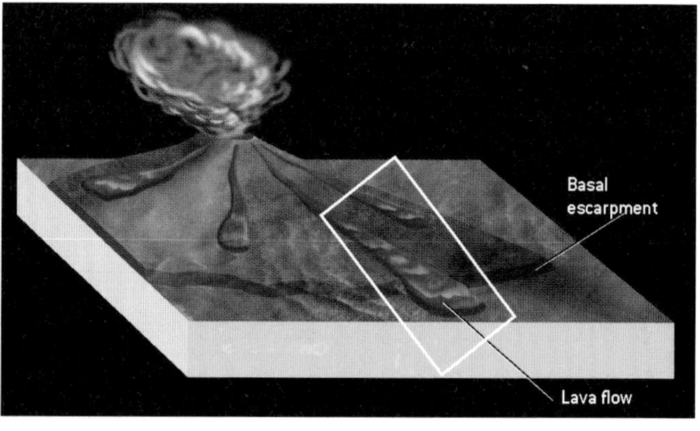

Lava flows on the slopes of Olympus Mons (above). High-resolution images from Mars Global Surveyor, Mars Odyssey and Mars Express have enabled scientists to work out the age of the most recent lava flows by counting small craters. Some of these are 100 million, some as little as 2 million years old: geologically, very young. It is possible that volcanic activity might still be occurring on Mars (illumination is from the left in the image). (Courtesy NASA, JPL, ASU.)

Accumulation of lava layers. Raised features several kilometres thick, composed of stacked lava or ash layers, are evidence of particularly active and only recently extinct volcanism on Mars. (Courtesy NASA, JPL, Malin Space Science Systems.) This region may be compared with the Deccan Plateau (above) in India with its famous 'traps', immense lava sheets which spread to a depth of 2 kilometres, formed 65 million years ago. (Courtesy Pour La Science.)

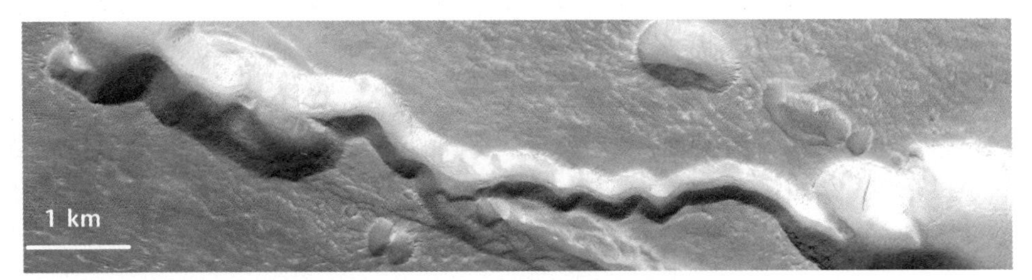

Example of a lava 'river' on the southern flank of the volcano Ascraeus Mons. The area shown in the image is 3 km across, with illumination from below. This 'river' has cut through multiple, generally shallow lava flows, lying upon other, more ancient ones. (Courtesy NASA, JPL, Malin Space Science Systems.)

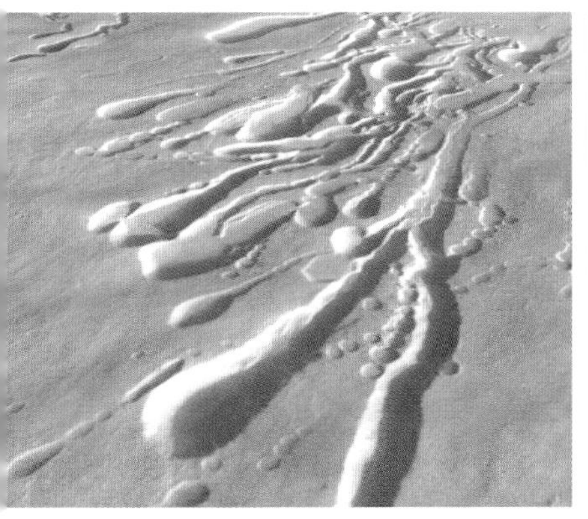

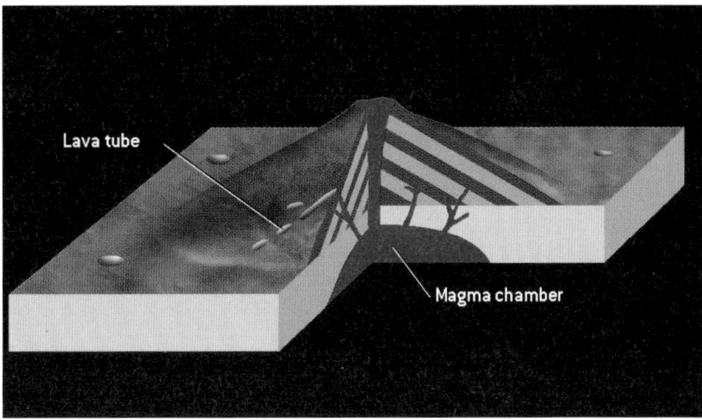

Collapsed lava tubes near Pavonis Mons, in the form of a line of crater-like pits (illumination from right). (Courtesy ESA, DLR, FU Berlin (G. Neukum).) As the flow cooled, lava deeper down continued to run beneath the cooling crust, forming a tube which eventually collapsed (see diagram).

plate-tectonic activity on Mars (see pages 27–29), together with its weaker gravity and low atmospheric pressure, meant that far greater amounts of gas, lava and ash were ejected.

Lava plains

There are also immense lava sheets, whose source is no longer evident, spread across plains such as Utopia, Acidalia and Lunae Planum. It is thought that this lava, which is particularly fluid, rose up through fissures and faults. In many ways, this kind of volcanism resembles that which characterises the lunar *maria*, the principal difference being that volcanic activity ceased on the Moon 3 billion years ago, while on Mars the lava still flowed comparatively recently. Naturally occurring sections though the slopes reveal that successive flows have accumulated to a depth of several kilometres. Since the end of the period of meteoritic bombardment, 3.8 billion years ago, some 60 million cubic kilometres of lava have been spilled across the Martian plains, at an average rate of 0.016 km^3 per year. The total volume of the volcanic formations (volcanoes and basalt plains) is about 500 million cubic kilometres.

5 Ridges and cracks on the surface

Relief sculpted by mechanisms from below

Until relatively recently, Mars was a geologically active planet. Features like the enormous volcanoes of Tharsis bear witness to this. Powerful convection currents within the mantle, and their interactions with the crust, have led to volcanism, and considerable reworking of the surface. Some of these surface deformations appear as extensions or spreading of the crust, while others are compressive in nature.

Tharsis is particularly interesting. In this region, the crust has been forced upwards and has been thickened by convergent convection currents below it. The upward motion caused stretching , and a radial pattern of faults spread from the centre of the upthrust. Between the faults, more often than not, are collapsed

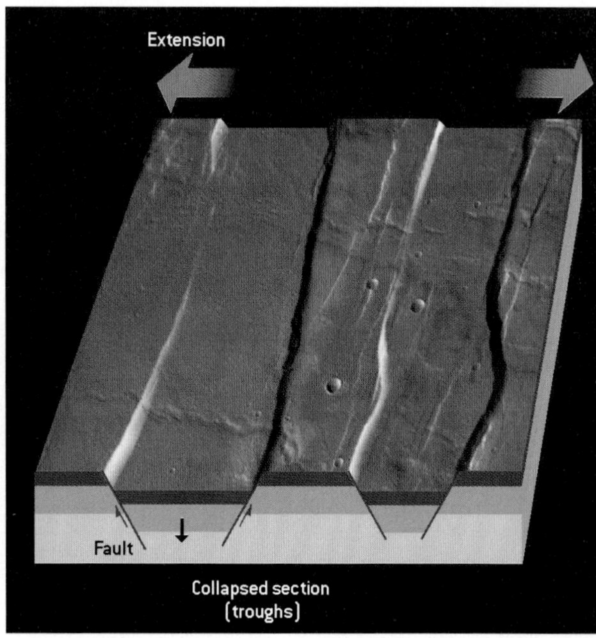

Graben. The image shows graben (parallel fault-bounded valleys) in the region of Acheron Fossae. They were created by the raising of the volcano Alba Patera. Stretching of the crust has considerably deformed a more ancient crater, 55 kilometres in diameter (Mars Express image, HRSC). (Courtesy ESA, DLR, FU Berlin (G. Neukum).)

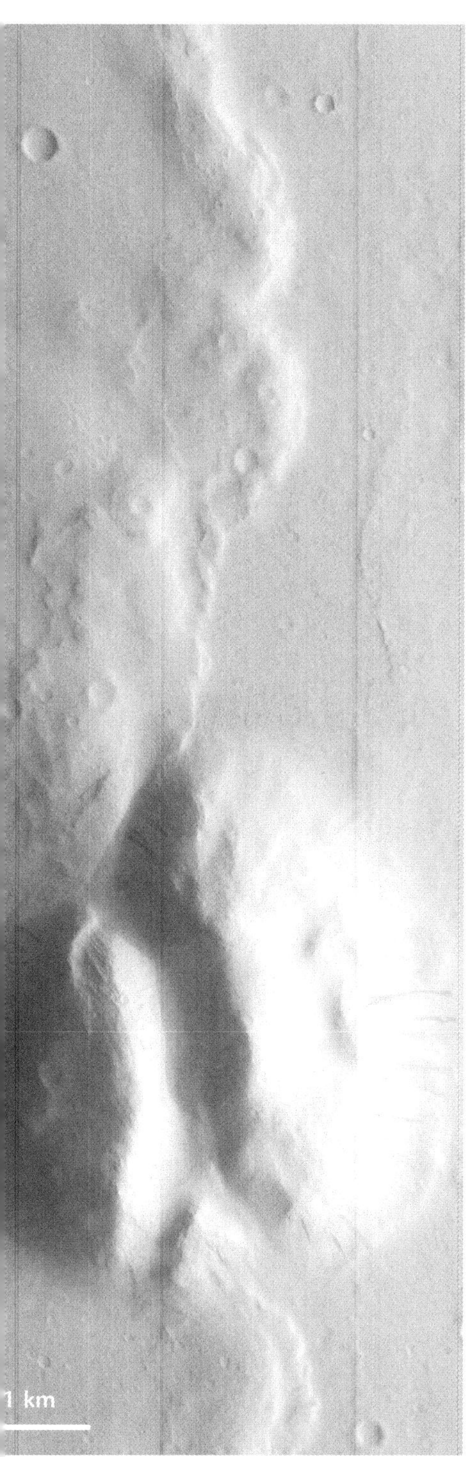

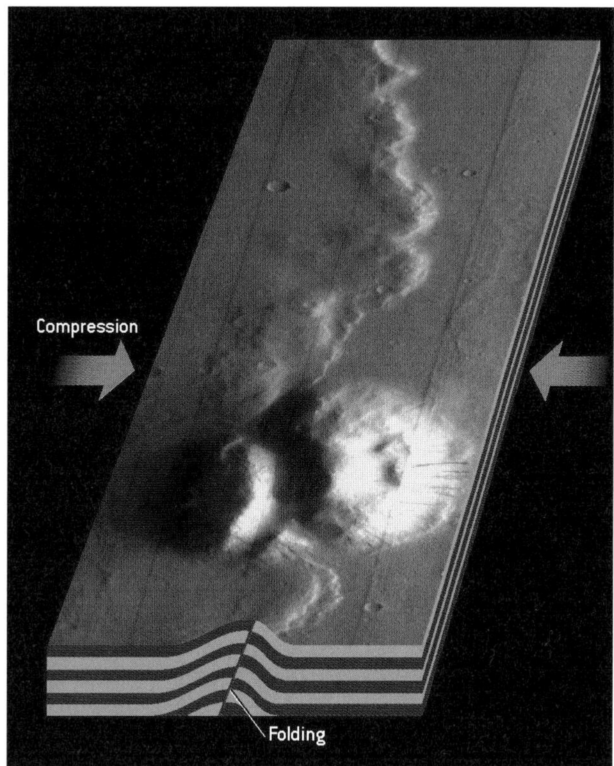

Wrinkle ridges. Ridges like these are found near Tharsis, in the Lunae Planum region (MOC image, with illumination from the right). (Courtesy NASA, JPL, Malin Space Science Systems.)

troughs (*graben*). Some compressive structures appear as ridges, arranged concentrically around the centre of the bulge.

Graben

These occur essentially on Tharsis. They can be several hundred kilometres long and a few kilometres wide. They are really sections of the surface which have slumped between the faults caused by the stretching of the crust.

Compression ('wrinkle') ridges

The ridges on Lunae Planum resemble those on the lunar *maria*. These linear landforms are hundreds of kilometres long. Their average width is 5 kilometres, and their maximum height is about 300 metres. It is likely that they result from compression of the surface. These compressive or wrinkle ridges are rooted in the crust to depths of a few kilometres. At these depths, the presence of volatile materials such as water, and hydric or carbonic ices, has doubtless favoured the sliding of one layer upon another. Other ridges may be signs of the contraction which took place in the Martian crust during the planet's cooling phase.

6 Valles Marineris: the valley of wonders

A rift 3500 km long and 7 km deep

Valles Marineris, a canyon system 3500 kilometres long (composite image, Viking). (Courtesy NASA, USGS.)

One of the most astonishing aspects of the planet Mars is the extreme nature of its relief. In fact, Mars boasts the biggest 'canyon' in the solar system: Valles Marineris. This canyon is 3500 kilometres long, 300 kilometres wide, and 7000 metres deep. Stretching eastwards from the volcanoes of Tharsis, and almost

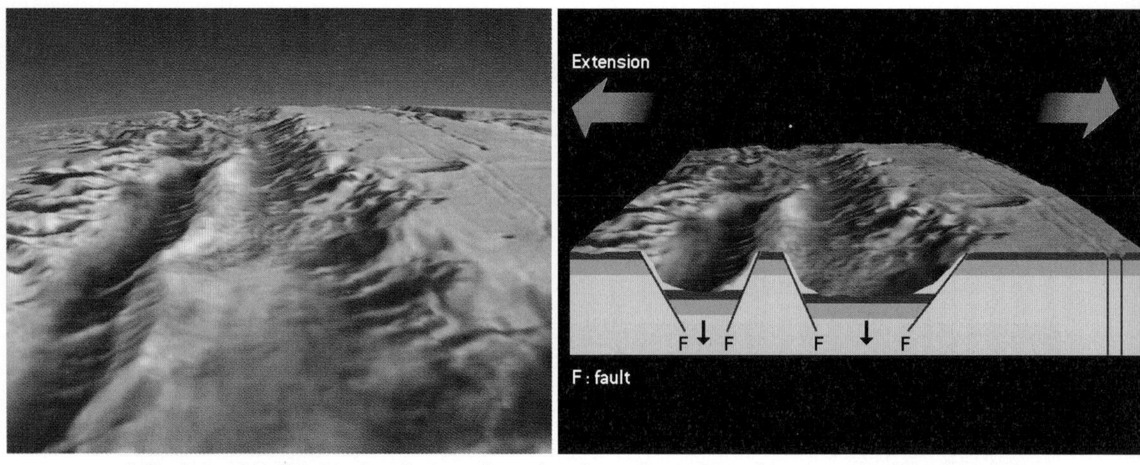

3-D view of Valles Marineris, based on data from the MOLA altimeter. The 'floor' of Valles Marineris has slumped along vertical fault lines. (Courtesy NASA, MOLA Science Team.)

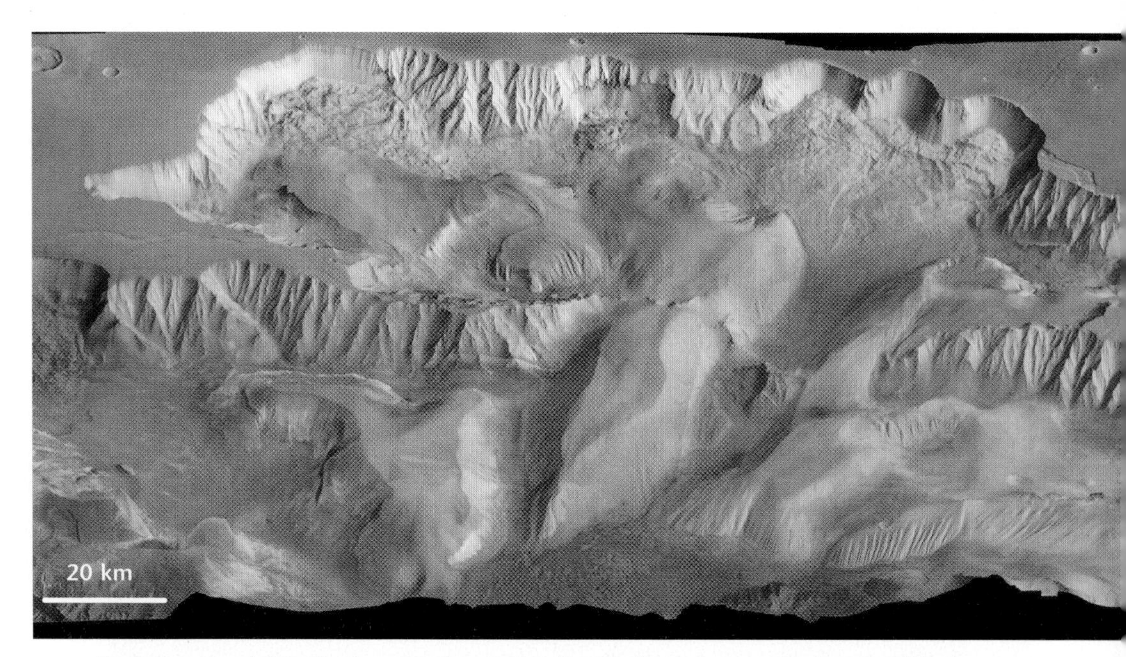

Enormous landslides in the central part of Valles Marineris (composite image, Viking). (Courtesy NASA.)

parallel to the equator, Valles Marineris opened out as Tharsis bulged upwards with the extension of Mars' crust (see previous pages).

This extension created a system of parallel faults along which lies a complex arrangement of collapsed trenches. Some planetologists see the Valles Marineris complex as a much larger version of the East African Rift Valley. On Mars, however, the plateau has opened and collapsed to a depth of 7000 metres. The stretching of the crust allowed magma to rise into the faults, and it later receded, causing localised collapsing of the surface. This phenomenon has left behind it the depressions, resembling lines of (sometimes adjoining) craters, found in the vicinity of collapsed trenches.

Landslides have enlarged the trenches of Valles Marineris. For example, in the central area, which is 5000 metres deep, the canyon widens out to form the vast elongated depressions known as 'chasmata'. They have very steep sides, and sometimes floor deposits that have slumped and spread out to distances as great as 200 kilometres. The size of these 'fans' has led scientists to estimate that the landslides from which they originate fell at speeds approaching 300 km/h. In places, the canyon floors have thick stratified deposits. These ancient features are of mysterious provenance. The OMEGA mapping spectrometer aboard Mars Express revealed that some of them are rich in sulphates, possibly formed as shallow lakes evaporated, or as a result of precipitation involving very salty water (see pages 59–61).

7 A permanently frozen sub-surface

Vast amounts of ice may lie trapped in deep permafrost

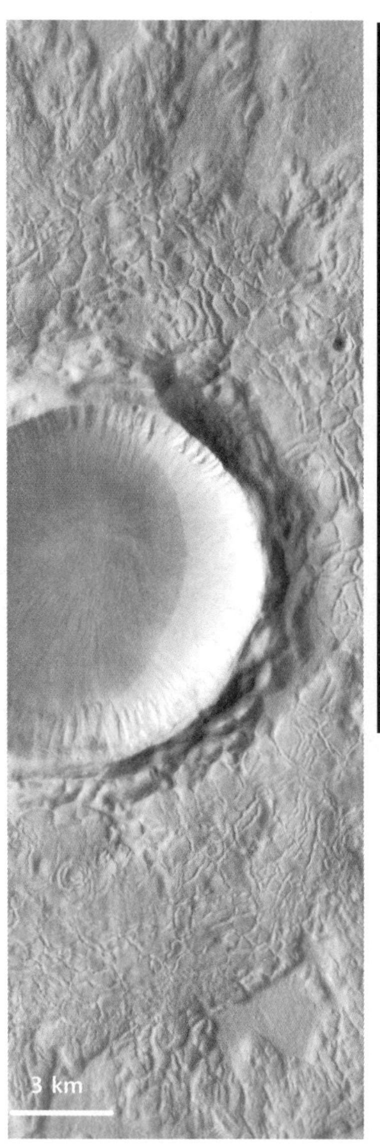

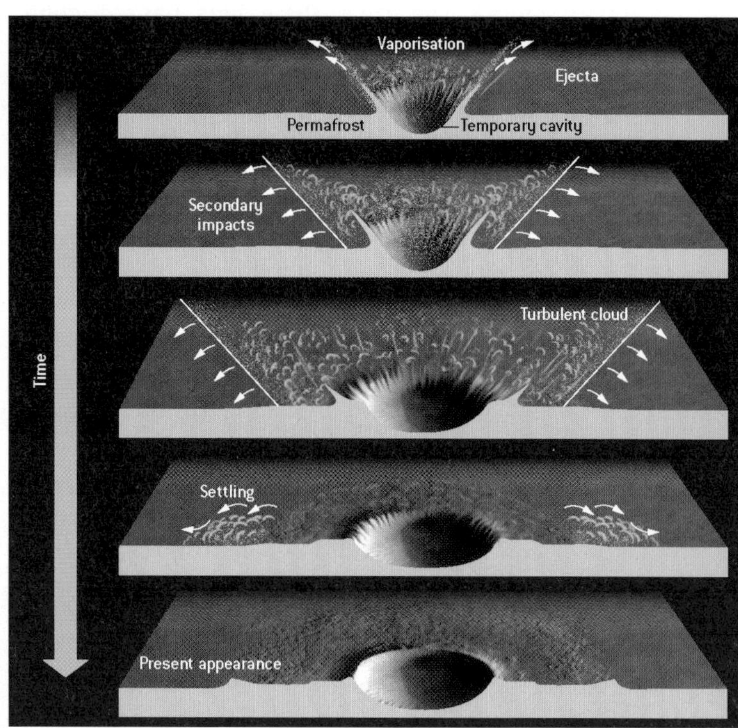

Craters in frozen ground. Lobate ejecta deposits around the impact crater (left) are probably due to the existence of sub-surface material rich in ice. The diameter of the crater is 10 kilometres. (Courtesy NASA, JPL, Malin Space Science Systems.) The diagram shows, firstly, the moment of impact, when a sudden rise in temperature causes the ice to melt and ejecta containing large quantities of water, ice and water vapour form a kind of *nuée ardente*. Then, having fallen back, the ejecta continue to slide across the surface for some distance, to form the lobes observed. Such 'lobate-ejecta' craters are valuable indicators of the characteristics and state of the sub-surface material at the time of impacts. Careful studies of their morphology can provide information about the depth and ice content of the 'permafrost'.

The mean annual temperature on Mars at the surface is –70°C: well below the freezing point of water. In spite of this, there are no stable deposits of ice anywhere on the planet except in its polar regions. In 'tropical' latitudes, frost sometimes forms before dawn, rapidly evaporating (or more properly, 'sublimating', as it passes directly from the solid to the gaseous state) as the temperature rises through the morning hours. As the Sun warms the surface, the layer of atmosphere nearest to the ground dries out to such an extent that ice is compelled to sublimate, even at negative temperatures. In winter, in mid-latitudes or on sunless slopes, ice may sometimes lie for weeks, but here again, it will become unstable at the surface as the spring warming begins. However, if the ice is relatively isolated from the atmosphere (for example, beneath a layer of dust), and if temperatures remain low enough, it may be maintained stably in the solid state. Indeed, water can theoretically be present on Mars above latitude 40°, in the form of ice in cavities not far below the surface.

On Earth, in places where conditions are similar to those described (for example in northern Siberia or in Canada), sub-surface material is frozen to depths of hundreds of metres: permafrost. Could there be ice-rich permafrost on Mars? There are numerous observations which tend to confirm this. While most of the solar system's impact craters are surrounded by radially ejected material, thousands of Martian craters display ejecta with a lobate appearance, reminiscent of mudflows. For geologists, these craters provide valuable evidence of the existence of ice-rich permafrost below the surface of Mars.

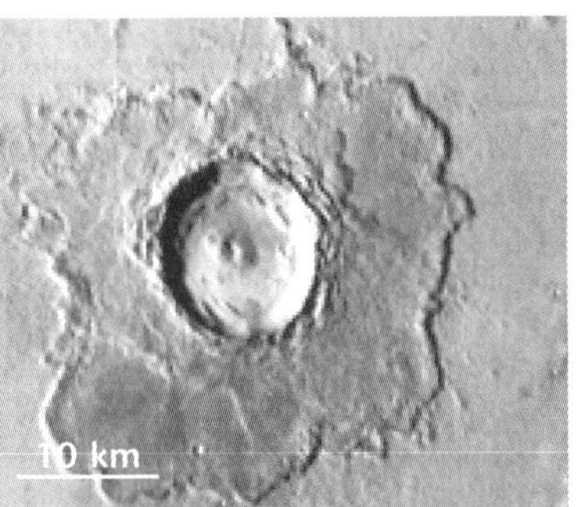

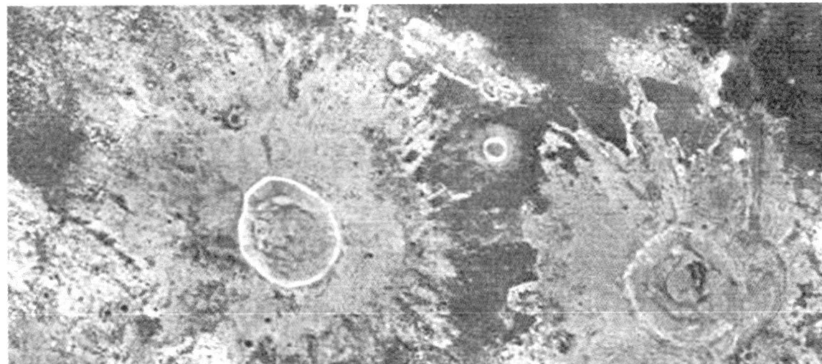

Lobate-ejecta craters as photographed by the Themis camera on Mars Odyssey, 2002. At left, a crater in the Chryse Planitia region, in visible light (i.e. human eye range). At right, craters in the Isidis Planitia region, photographed at night with the camera's thermal imaging system. The particular nature of each crater is revealed by temperature differences linked to variations in the composition and texture of the ejecta materials. On the plains of the northern hemisphere, the ejecta around this type of crater are quite extensive, suggesting the presence of considerable amounts of water ice at the moment of impact. Even the smallest craters have their lobate ejecta, confirmation of a soil rich in ice quite close to the surface. This concentration in the northern plains is probably evidence of an unequal distribution of ice in the sub-surface of Mars. Outflow channels and the hypothetical oceans (see pages 65–67) have surely contributed to the formation of well-developed ice-rich permafrost across the northern hemisphere. (Courtesy NASA, JPL, ASU.)

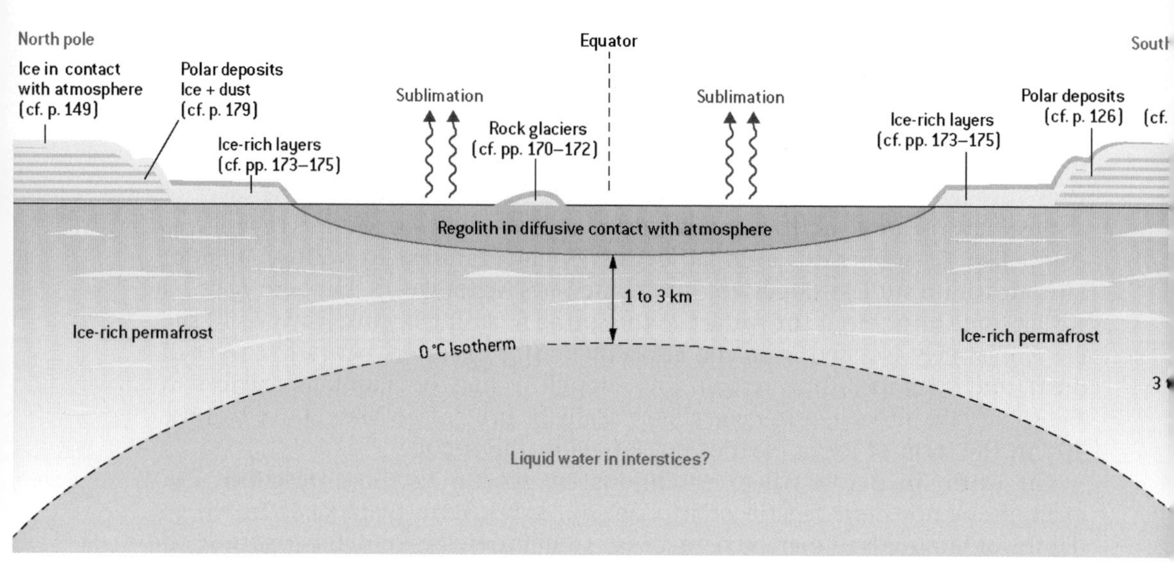

Water beneath the surface of Mars. Above the porous and ice-rich soil which is the permafrost, several kinds of much more recent icy deposits have been identified. The ice has mostly been transported in the atmosphere, deposited, and then insulated and protected by a layer of sediments (except near the poles, where the ice is exposed). The history of these deposits will be discussed in part 4, which deals with the Martian climate. The global permafrost begins at an estimated depth of between 300 metres and 1 kilometre in the equatorial zone, and between 300 metres and less than 10 metres in mid- and high latitudes, where average temperatures below –70°C allow the permafrost to stabilise near the surface. It is thought that the depth of the permafrost is between 1 and 3 kilometres in the equatorial zone, and between 3 and 7 kilometres in higher latitudes. Below these depths, the pressure of the geological layers, and geothermal heating from the interior of the planet, mean that water would generally exist in the liquid state.

8 The mystery of the chaotic terrains

What became of the 'missing material' of these terrains?

On the equator of Mars, the eastern end of the Valles Marineris canyon system opens out abruptly, continuing as a vast topographical depression whose surface area exceeds that of France. In this sector, the cratered plateau is cut into, to depths of nearly 3 kilometres, by jumbled ('chaotic') terrains. Some of these take the form of huge closed depressions, many hundreds of kilometres long. They are of mysterious origin. Vast, deep outflow valleys are closely associated with these terrains (see page 111).

A good example is seen in the region of Hydaspis Chaos, at the mouth of Valles Marineris. Here, the chaotic terrains are dotted with large numbers of pyramidal buttes about one kilometre high, situated along orthogonal faults and regularly spaced (see photo below).

How are such terrains formed? One of the principal mysteries is the absence of any outlet or any kind of gap through which material initially present may have been evacuated.

A commonly supported hypothesis suggests that this region of the cratered plateau was originally subject to permafrost. As the sub-surface material became warmed due to volcanic activity or climatic variations, the permafrost melted, causing the terrain to collapse in places.

Chaotic terrain: Hydaspis Chaos. The numerous hills (buttes) are the result of erosion involving melting permafrost. Illumination is from the top. (Courtesy NASA, JPL, ASU.)

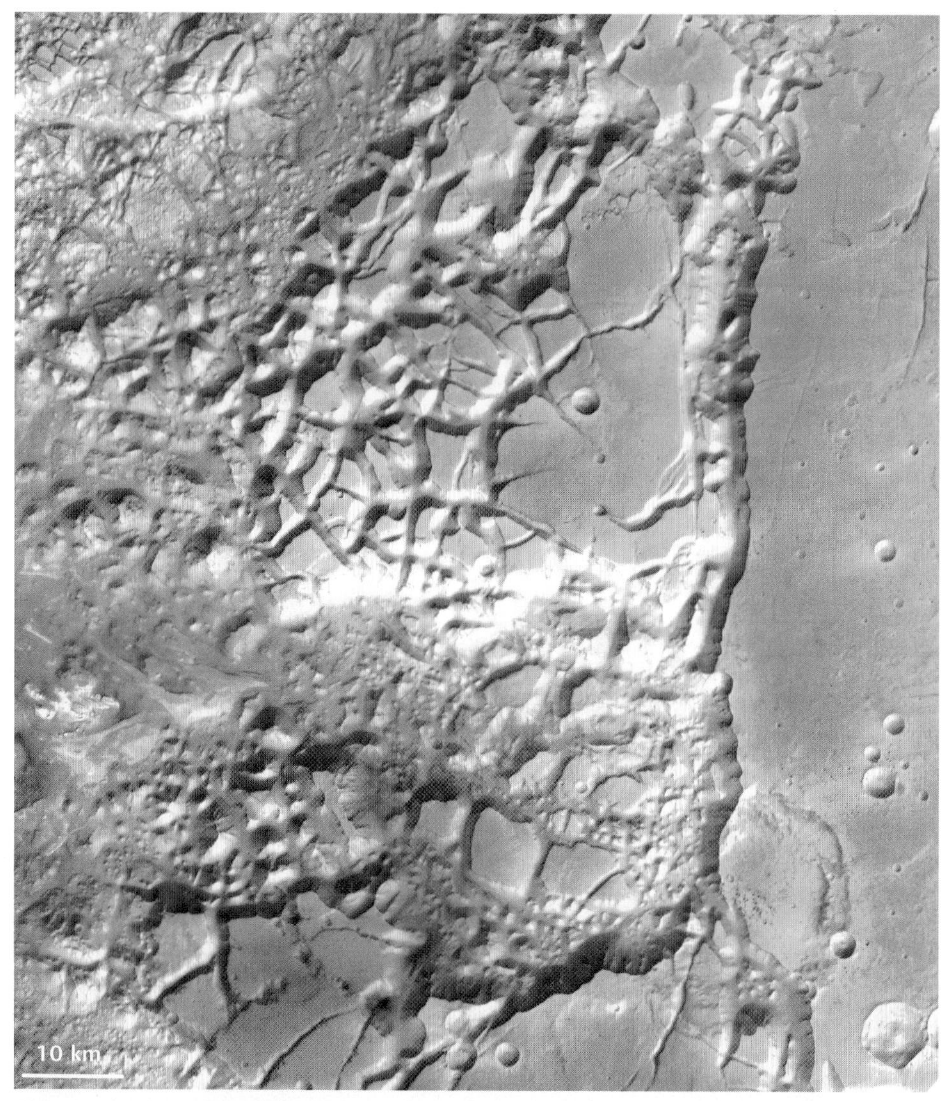

Chaotic terrain in the region of Aureum Chaos, south of Ares Vallis. The cratered plateau has been dissected and eroded to a depth of several hundred metres (HRSC camera, Mars Express). (Courtesy ESA, DLR, FU Berlin (G. Neukum).)

Another question remains: where is all the solid material (sediments and rocky debris) which was carried away during this thaw? One explanation might be that, in the permafrost in this area, ice was particularly abundant while rocky matter was scarce. When the ice melted, the volume would have been considerably reduced, and the land simply collapsed. The total amount of water 'excavated' from these chaotic terrains, if the ice-rich permafrost theory is correct, would represent a planet-wide layer of water 40 metres deep.

9 Outflow channels

Great valleys carved out by cataclysmic floods

To the north-east of Valles Marineris, immense, deep valleys cut through the cratered plateau and extend towards the plain of Chryse Planitia (see map below). When they were discovered, there was considerable excitement: it

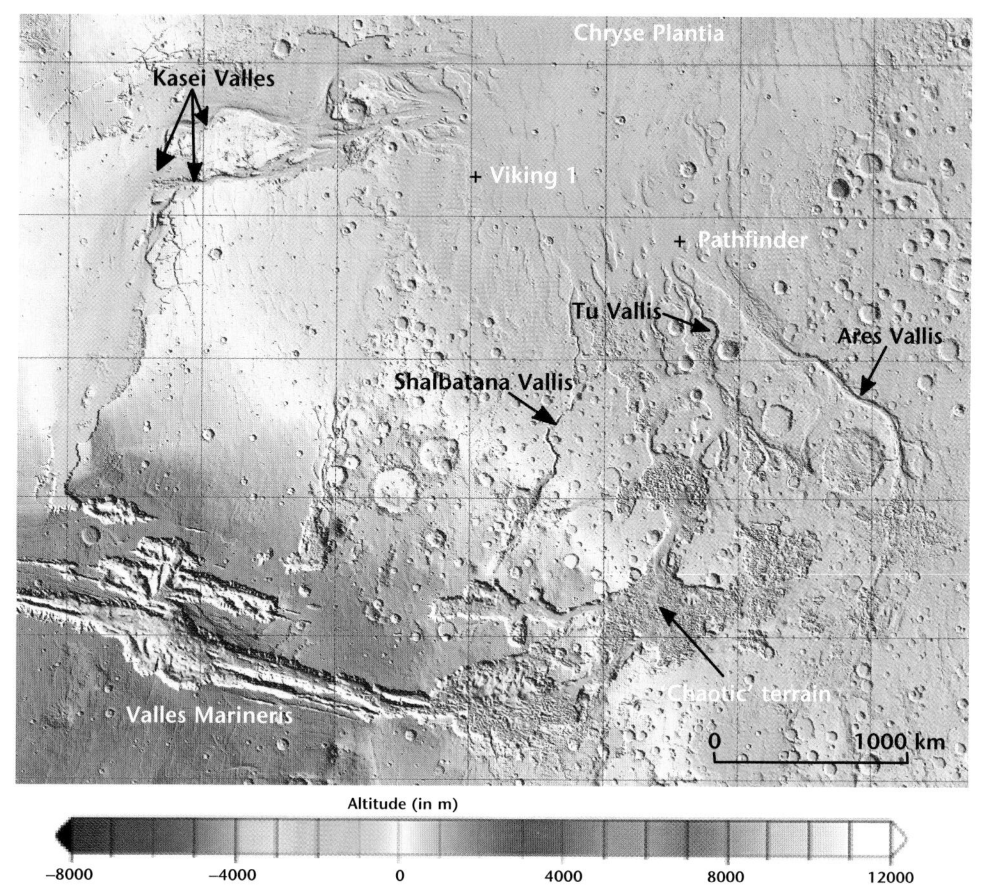

Topography of the Chryse Planitia area, where the main outflow channels debouch (black arrows). Most of these channels are between 5 and 25 kilometres wide, and 2 kilometres deep. They may be up to 1500 kilometres long. Their longitudinal slopes are very shallow. (Courtesy NASA, JPL, ASU.)

seemed highly likely that they had been fashioned by flowing water or mud! Their origins and morphology seemed, however, to have little in common with those of the fluvial valley networks on the cratered plateau (as described on pages 50–55). Some of these features appeared to be quite recent on the geological timescale. These outflow channels show a peculiar morphology: they have no tributaries, and their geographical origin is quite localised. On the floors of these channels, large numbers of channels sweep around prominent obstacles such as impact craters, sculpting teardrop-shaped structures tapering in the direction of the flow. The geometry of these streamlined islands has led researchers to conclude that they were carved out by very turbulent running water, rather than by ice, lava or mud.

Geomorphologists believe that only sudden, short-lived bursts of vast amounts of water can explain the characteristics of the outflow channels. It is possible to estimate the amount of water involved during these catastrophic episodes by studying the widths of the channels. To take the example of Ares Vallis, which is 25 kilometres across: if we imagine a layer of water ten metres deep, then the peak discharge rate could have been greater than 100 000 cubic metres per second. Other outflow channels on Mars might have experienced flows of about 70 million cubic metres per second, which is more than two thousand times that of the Mississippi River! Where did all this water come from? Several scenarios have been suggested.

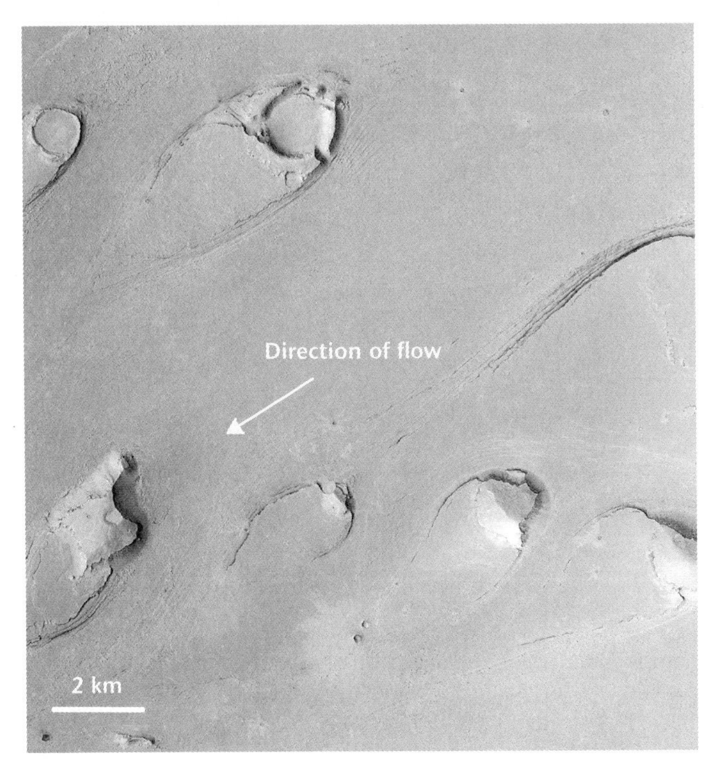

The outflow channel Athabasca Vallis, in the Cerberus region, not far from Elysium. In this image, we see streamlining of features in the direction of the flow. Counting impact craters (see 'Chronology of Mars', pages 12–14) suggests that the flow occurred less than 10 million years ago (Mars Global Surveyor). (Courtesy NASA, JPL, Malin Space Science Systems.)

The outflow channel Ares Vallis. Here, 100 000 m^3 of water per second could have flowed, after the sudden outburst of 10^{13} m^3 of water from an underground reservoir under pressure within the permafrost. This image of Ares Vallis was taken in 2002 by the Themis camera aboard Mars Odyssey. Illumination is from the left. (Courtesy NASA, JPL, ASU.)

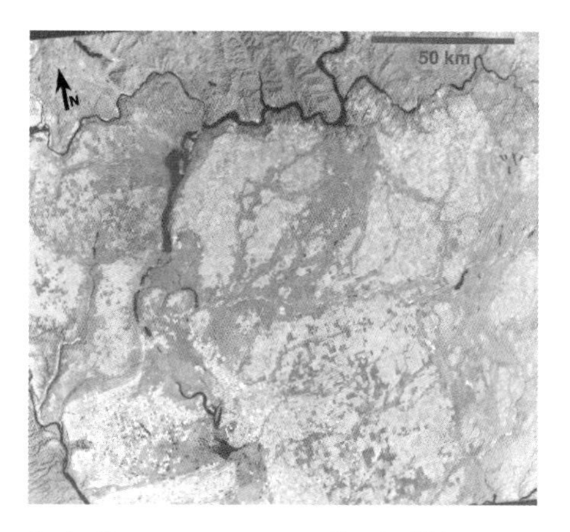

A similar occurrence on Earth. This Landsat image shows the Scabland area in Washington state (USA). Hydrologist Dr Victor Baker, of the University of Arizona at Tucson, compared this area with Martian outflow channels, characterised by few tributaries and not very sinuous courses. He advanced the hypothesis of a catastrophic origin, showing that the Scabland area experienced tremendous flooding after the discharge of an ancient glacial lake in the Pleistocene era (1.8 million to 11 000 years before the present). Colossal volumes of water, suddenly liberated, carved huge, non-sinuous valleys with smooth floors, comparable in size to Martian valleys.

Mike Carr, of the US Geological Survey (California), proposes that the floods associated with the outflow channels to the north of Valles Marineris represent the sudden liberation of an underground water reservoir, trapped under pressure within the Martian permafrost. This reservoir would have been capable of supplying almost 10^{13} (10 000 billion) cubic metres of water.

In some other channels, for example Athabasca Vallis (see image on facing page), the water seems to have come from the melting of permafrost due to lava rising through fractures in the crust.

10 A drive around the mouth of Ares Vallis

In 1997, the Pathfinder mission touched down on the floor of an outflow channel

What do outflow channels look like, as seen from within? In July 1997, the Pathfinder lander, carrying its little robot rover Sojourner, transmitted a panorama of the Ares Vallis outflow channel (see page 111). Jumbled rocks, tens of centimetres across, lay everywhere, among sparse dunes. The landscape differed little from that photographed by Viking 1, which had touched down in 1976 some 800 kilometres to the north-west.

Area at the mouth of Ares Vallis, photographed by the Pathfinder probe. The 30-cm high rover Sojourner examines a rock known as 'Yogi'. On the horizon, the 'Twin Peaks', hills with sedimentary/volcanic strata. In the foreground are the solar panels of the probe. (Courtesy NASA, JPL.)

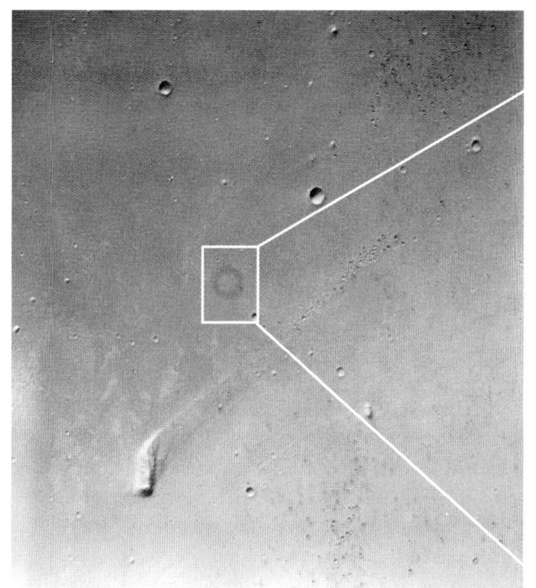

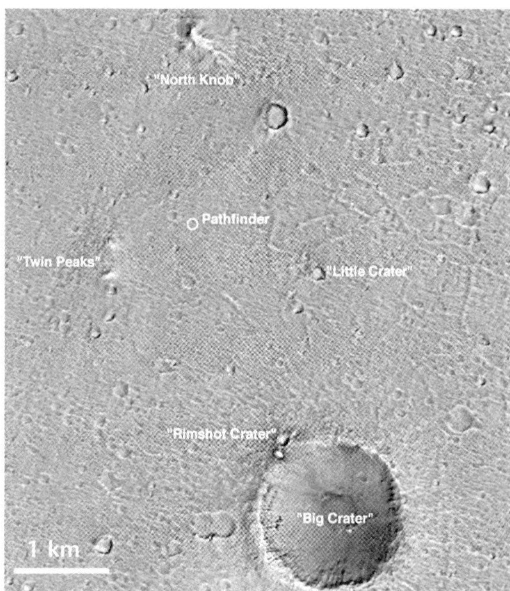

The Pathfinder landing site. This large-scale photograph (left) from the Viking orbiter (1976) shows the Pathfinder landing site on a smooth, wide plain without major topographical features. Traces left by the passage of water are visible: a butte at the origin of a streamlined landform seems once to have been in the midst of a flood. On the high-resolution image (right) from Mars Global Surveyor, we see aligned dunes and an impact crater several kilometres in diameter. (Images courtesy NASA, JPL, Malin Space Science Systems.)

Where did the rocks come from?

Had they been transported here by the enormous floods of water or mud which, it is believed, swept through the channels ages ago? Did they represent ejecta from impact craters in the vicinity? Or might they be the remnants of the original local rocks, worn away by erosion? All three hypotheses are possible. Now, some of these blocks are triangular and are marked by striations, implying that intense wind erosion has been at work (see pages 134–136). Others bear sharp rectilinear ridges, suggesting that they are among the ejecta from violent meteoritic impacts. Several of the blocks in the image overlap, or lie at angles, inviting the conclusion that a flowing liquid may have left them there.

Why is the ground orange?

The whole surface of Mars reflects red light preferentially, and absorbs violet and blue light. The planet's colour is echoed in the names given to it in many languages. In Ancient Egypt, Mars was *har decher*, 'the Red One'; in India, *Lohitanga* the 'Red-Bodied'; to the Romans, Mars, god of war, was associated with blood.

Chemical analysis. Analyses performed *in situ* with a spectrometer on Sojourner show that these blocks are mostly of the andesite type, resulting from past volcanic activity (blue arrows). All these blocks are covered with a thin layer of dust a few micrometres thick and rich in magnetised iron oxides (magnetites – red arrow). The soil has a thin crust rich in magnesium sulphate. Such a composition tends to suggest the presence of mineral salts, probably left behind after the evaporation of water in the past. (Courtesy NASA, JPL.)

Terrestrial volcanic blocks in Death Valley, California. The biggest of these blocks measures about 80 cm. These are similar to the rocks scattered about the surface of Mars, and are dark in colour, with numerous vacuoles. (Courtesy F. Costard.)

Scientists have long seen Mars as a 'rusty' planet, given its colour, and data from Pathfinder showed that the soil is indeed about 17–18 percent ferric oxide (Fe_2O_3). Ferric oxide is in fact the second most abundant constituent of the Martian soil, after silica (silicon dioxide, SiO_2, at nearly 43 percent). Twenty years before, Vikings 1 and 2 had registered almost identical soil compositions at their respective sites, demonstrating that the surface of Mars is covered by a fine layer of uniformly spread dust, moved around by dust storms over billions of years. Why is there so much iron on Mars? Analysis of the SNC meteorites from Mars (see pages 30–32) tells us that, generally, its mantle has two and a half times more iron than is present in the Earth's mantle. The probability is that Mars' weaker gravity has caused the process of differentiation (see pages 24–26) to be less efficient.

11 A second Siberia

Scientific inspiration from the East

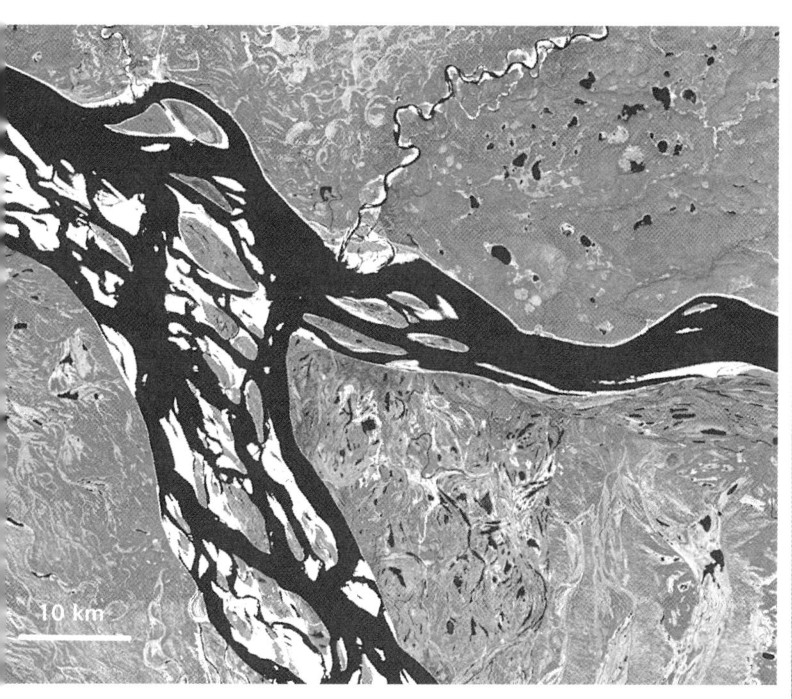

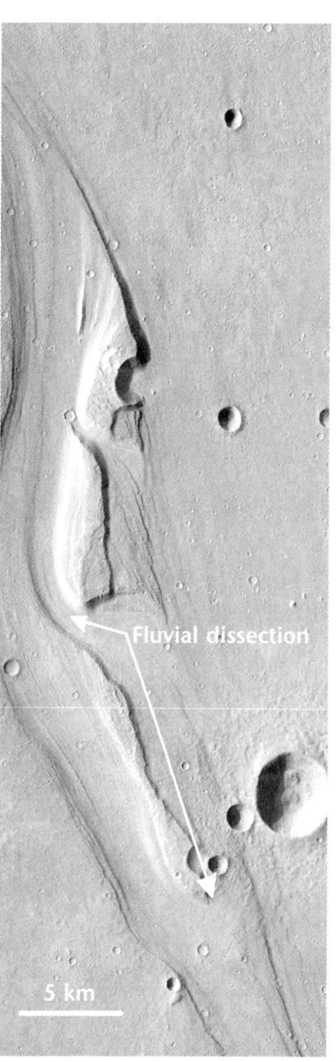

The valley of the River Lena, and Martian outflow channels. The Lena (left), 4000 kilometres long, has a flood plain 25 kilometres wide below Yakutsk, capital of the Yakutia republic in Siberia. (Courtesy NASA, Landsat.) The bed of this river has many features in common with the outflow channels of Mars. Both have developed above permafrost, and they are of identical scale, with low longitudinal slopes and few tributaries. Violent flow episodes in the Siberian spring suggest what must have happened on Mars. Because of the great volumes of water involved, these floods are called 'catastrophic'. The flow of the Lena attains 50 000 cubic metres per second: the value for the flooding in Ares Vallis is thought to have been as much as 100 000 cubic metres per second.

The junction of the Martian valleys Simud and Shalbatana (right) is characterised by multiple channels. Upstream in certain valleys such as Simud, channels are sub-divided and form streamlined islands . These features are quite similar to those in the Siberian flow valleys in central Yakutia. The multiple channels are caused by very strong variations in flow from one season, or one month, to the next. Very wide valleys with flat, shallow channels are the result (Themis, Mars Odyssey). (Courtesy NASA, JPL, ASU.)

Planet Mars is similar in some ways to Siberia. Temperatures can fall to –100°C during Mars' equatorial winter, and rise to +23°C in summer. In central Siberia, it is not unusual to encounter winter temperatures of –70°C and summer temperatures of up to +30°C. In these extreme environments, low temperatures are responsible for deep and continuous permafrost, which plays an essential part in the evolution of the landscape. It forms a veritable carapace, said to be 'as hard as concrete'. In central Yakutia (the Siberian Sakha Republic), the depth of the permafrost may attain 1500 metres. On Mars, we may find similar values (see pages 106–108).

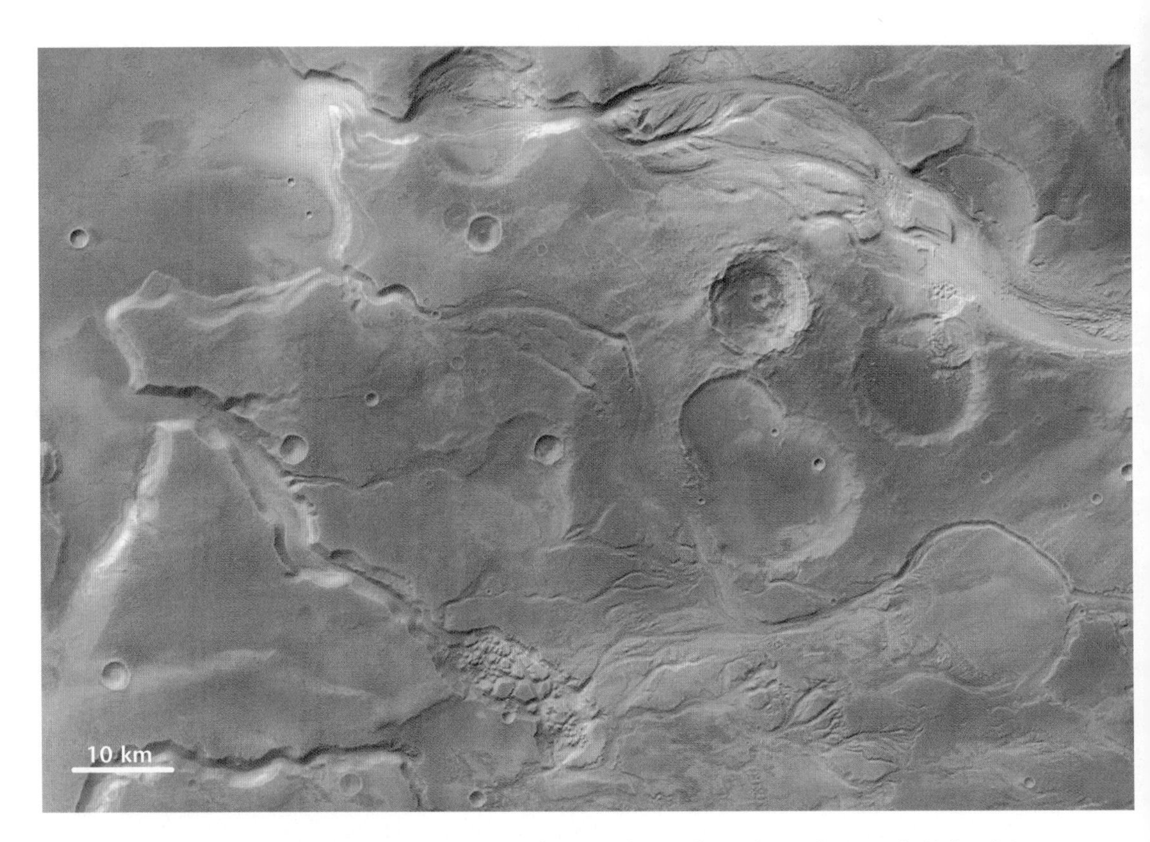

Patterns of erosion in the frozen soil of Mars. The outflow channel Mangala Valles (Mars Express, HRSC). (Courtesy ESA, DLR, FU Berlin (G. Neukum).)

Rapid erosion in frozen banks. In Siberia in spring the thaw of the river and the melting of snow combine to create a tremendous outflow capable of cutting back river banks by 30 metres over a few months. This erosion exposes massive layers of ice in the permafrost (arrowed). This 25-metre cliff is on the river Lena. Planetologists believe that layers similar to these exist in the plains of the northern hemisphere of Mars. (Courtesy F. Costard.)

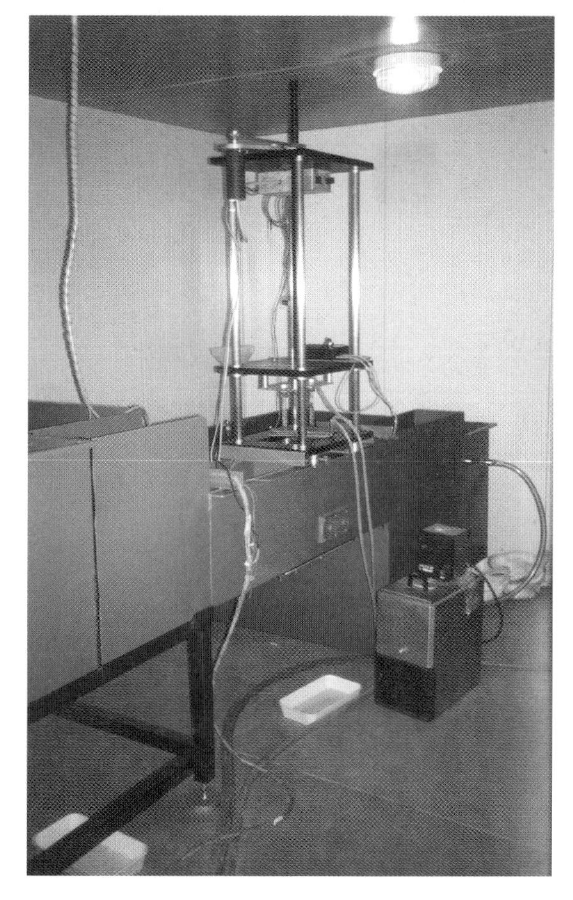

Permafrost in the laboratory. In order to study in detail the exact conditions in which certain processes on Mars and in Siberia occurred, planetologists try to reproduce those conditions in the laboratory. The photo shows a hydraulic channel set up in a cold room by researchers at the University of Orsay, with a view to studying the processes of thermal erosion in frozen soil, such as occur during flow episodes on Mars. (Courtesy F. Costard.)

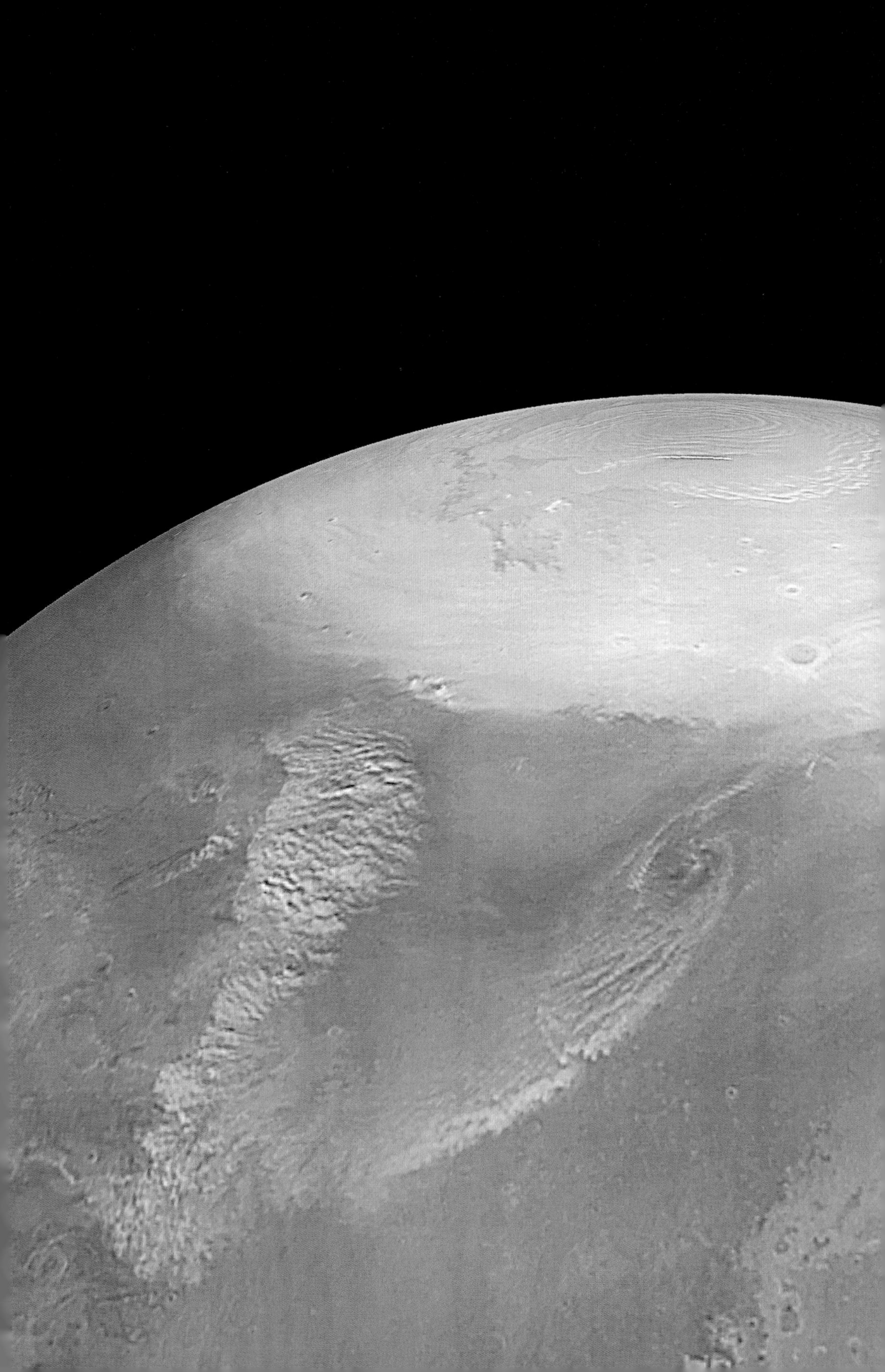

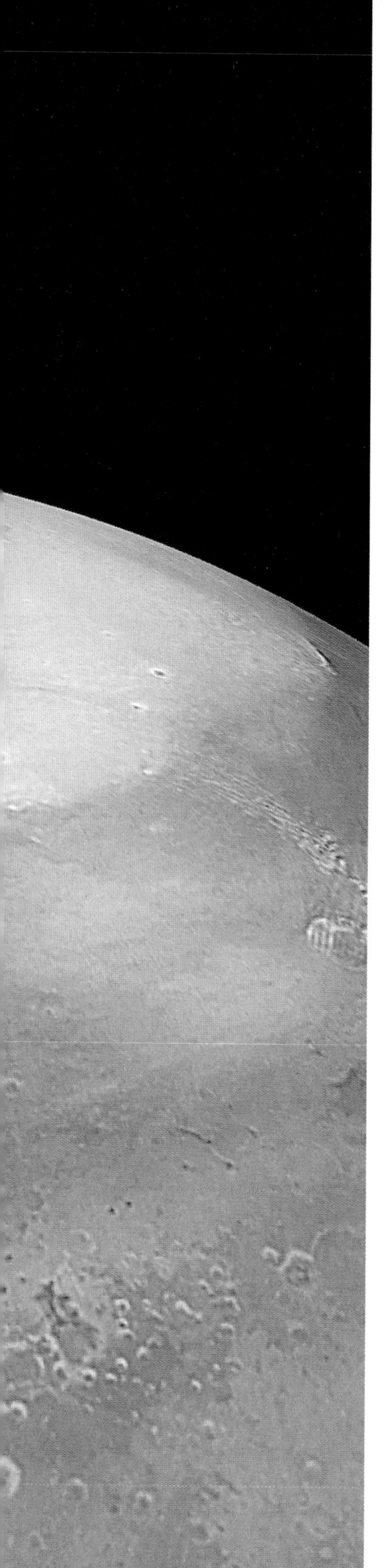

Climates and storms

Mars today

Now, returning to the present, a look at current Martian meteorology, with its storms, dust devils, snowfall, trade winds and anticyclones. . .
Day after day, atmospheric movements mould the surface of Mars, at the whim of changing weather. The climate of the red planet is not fixed, and as the millennia pass, it is subject to slow and far-reaching fluctuations brought about by orbital variations.

Spring comes to the northern hemisphere of Mars (Mars Global Surveyor, May 2002). The north pole is still covered by a cloak of carbon dioxide snow, laid down during the northern winter. Further south, mid-latitude depressions cause strong winds, which raise fine mineral dust. Dust storms are raging beneath the whitish wreaths of ice-crystal mists. (Courtesy NASA, JPL, Malin Space Science Systems.)

1 Seasons and temperatures

A 'hyper-continental' climate

Mars, as it is reveals itself to us today, is a planet of changes. They are driven by the seasons and by the vagaries of its climate. Since the eighteenth century, astronomers have been able to watch the shrinking of the Martian polar caps at the onset of spring, the shifting patterns of Mars' clouds, and even colour changes on its surface. These phenomena came as no surprise: they merely

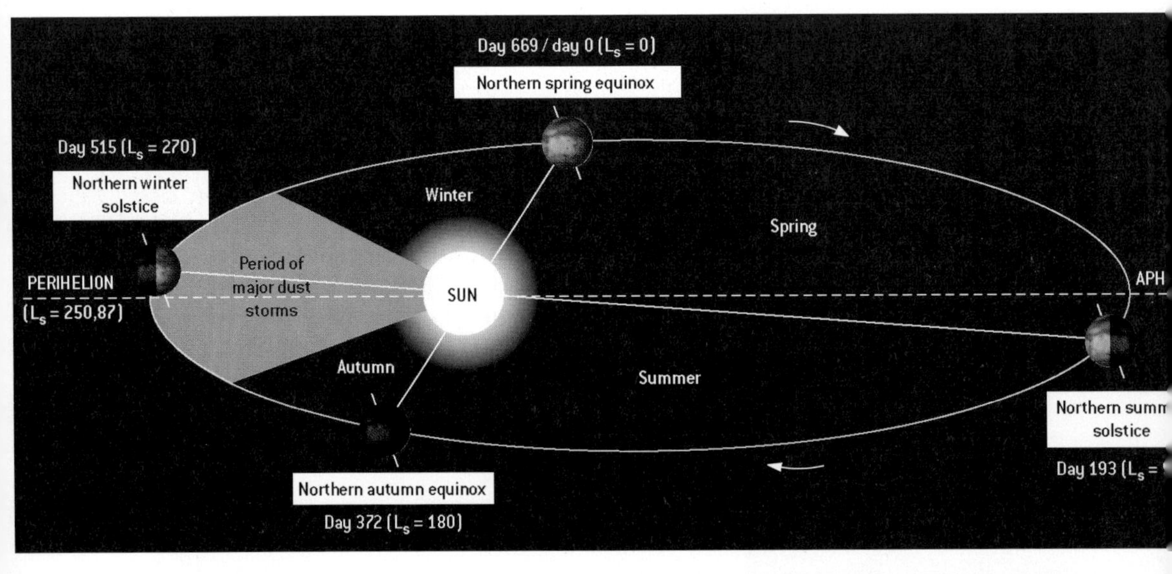

A Martian year. Although Mars, like planet Earth, has a cycle of seasons, its orbit is more eccentric and its year longer than Earth's. At perihelion (the closest point to the Sun in Mars' orbit), which occurs each year at the end of the southern spring, the planet is approximately 20% nearer to the Sun than in the opposite season. This state of affairs means that Mars then receives almost 50% more solar energy. The velocity of Mars is at its greatest at perihelion, causing seasons to be of very unequal length. The southern summer is warm, but short. On the diagram, the dates and seasons are shown with reference to solar longitude (L_S), the position of Mars relative to the Sun measured in degrees from the vernal equinox in the northern hemisphere. In the absence of a calendrical system like Earth's, the solar longitude serves to specify Martian dates and seasons throughout the year. In the diagram above the seasons described are for the northern hemisphere of Mars. In the southern hemisphere, the seasons are reversed: northern winter = southern summer; northern spring = southern autumn; northern summer = southern winter; and northern autumn = southern spring.

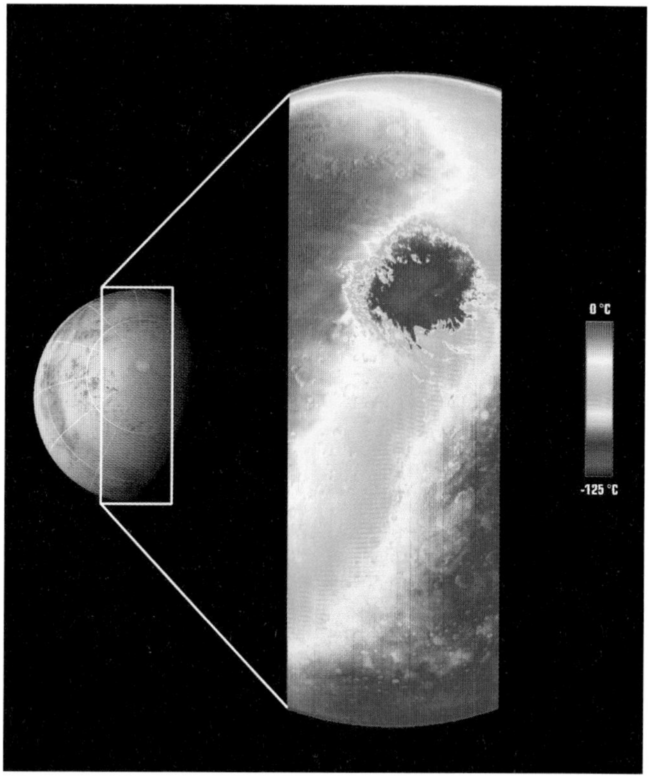

Temperatures at the surface. A temperature map for Mars' southern hemisphere, shortly before the summer solstice (based on data from the Themis infrared camera, Mars Odyssey). The contrast in temperature between day (red) and night (blue) is evident. At the south pole, a cap of carbon dioxide snow at –125°C, laid down during the preceding winter, is still visible (see pages 155–160). (Courtesy NASA, JPL, Arizona State University.)

reinforced the belief that all planets were made in the image of the Earth. Furthermore, the observers noticed that Mars rotates on its axis in a period similar to that of the Earth's day (a day on Mars lasting 24 hours and 40 minutes). Also, both planets' equatorial planes are inclined at a similar angle (obliquity) to the planes of their orbits (25.2° for Mars, 23.4° for Earth). Mars, like the Earth, was seen to have four seasons, although the cycle of seasons (one Martian year) lasts for 669 Martian days, or 687 Earth days. Alas for those astronomers of old, these resemblances do not make Mars 'another Earth'.

On a fine afternoon on Mars, in southern summer, the weather may be quite mild: 20°C, with a gentle trade wind... But with the night come glacial conditions, as temperatures fall by tens of degrees to –100°C, and remain there until morning (see temperature chart on page 124). Martian soil is dry and granular, and inefficient at storing heat: its 'thermal inertia' is very low compared with the surface of the Earth, with its oceans. Also, since the atmosphere of Mars is so tenuous, variations are much more marked. Briefly, the climate of Mars is 'hyper-continental'.

One result of this is that the cycle of seasons is much more marked on Mars than on Earth, especially at the two solstices. At these times, both planets are inclined in such a way as to present one pole towards the Sun all day: so the 'summer pole' receives the most solar energy throughout the day. On Earth, the thermal inertia of ice, and of the oceans, and especially the reflection of radiation from ice and snow, ensure that the North Pole is not hotter than, for example,

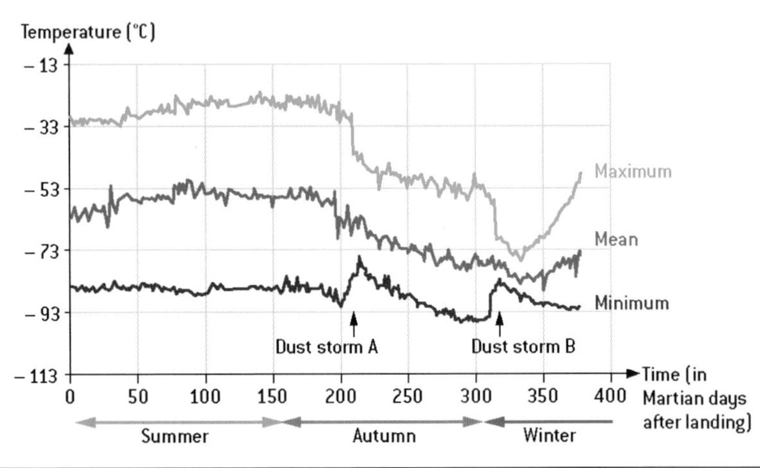

Viking 1, the first extraterrestrial weather station. The temperature of Mars, measured by Viking and transmitted directly from Chryse Planitia (22° N: see topographic map of Mars on page 111). A thermometer, an anemometer and a barometer are located at the end of the arm on the right. These instruments monitored temperature, wind speed and pressure at 1.6 metres from the ground from July 1976 until August 1979. From the beginning of northern autumn onwards (see diagram), the mean temperature slowly decreased until the winter solstice. The sudden onset of two global dust storms greatly lessened the amplitude of the temperature variations during the day. By day, the suspended dust screened out the Sun, and by night, it radiated in the infrared and limited the cooling of the surface. This mimics the effect of clouds on Earth: temperatures are milder when the sky is overcast than when the sky is clear at night. (Courtesy NASA.)

Death Valley in California. These effects are negligible on Mars, so temperatures (averaged out over the day) are at their highest in the polar region of the summer hemisphere at the solstice. As one moves from the summer pole towards the winter pole, temperatures continuously decrease. At the Martian equinoxes, however, we encounter a situation more familiar to us: cold polar regions, and warmth around the equator.

2 The atmosphere of Mars

A thin envelope of almost pure carbon dioxide

The Martian atmosphere above the Argyre Basin, photographed by Viking. (Courtesy NASA.)

Until the 1960s, it was believed that the atmosphere of Mars consisted of nitrogen, not very different from that in the Earth's atmosphere. Astronomers based their opinions on precise measurements, but had not taken into account the presence of the dust which creates the optical illusion that the Martian atmosphere is dense. They were therefore taken aback when, in 1965, Mariner 4 transmitted signals through that atmosphere as it passed behind the planet, revealing that it was tenuous, and that the mean surface pressure on Mars was around 6 hectopascals (the value on Earth at an altitude of about 30 kilometres)! Mars' atmosphere consists of 95% carbon dioxide, and would be toxic to

Gas	Symbol	Mars	Earth
Carbon dioxide	CO_2	95.32%	0.035%
(Molecular) nitrogen	N_2	2.7%	78%
Argon	Ar	1.6%	0.93%
(Molecular) oxygen	O_2	0.13%	20.6%
Carbon monoxide	CO	0.07%	0.00002%
Water	H_2O	~0.03%	~0.4%

Principal constituents of the atmospheres of Mars and the Earth.

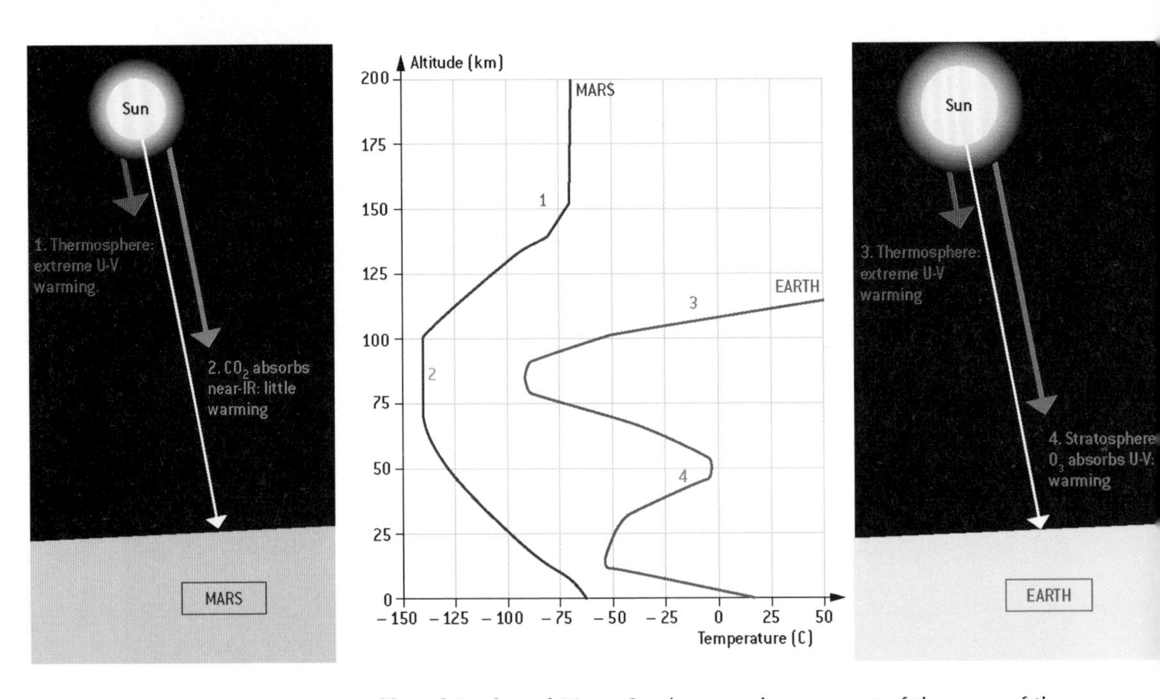

Mean temperature profiles of Earth and Mars. On the two planets, most of the mass of the atmosphere is confined to the lowest few kilometres. In this well-mixed layer, the troposphere, temperature falls off rapidly with increasing altitude. Between 20 and 60 kilometres, the thermal structure of Earth's atmosphere is characterised by a thick layer of ozone warmed by the absorption of ultraviolet radiation. Nothing like this occurs on Mars, where the ozone layer is negligible. Instead, between 60 and 120 kilometres up, the thermal profile stabilises due to the absorption of solar radiation in the near infrared by CO_2. Above 120 kilometres, in the thermospheres of both Mars and the Earth, the few molecules present are exposed to the most energetic photons of extreme ultraviolet radiation, which dissociates them and warms them considerably. The temperature of the thermosphere shows strong daily and seasonal variations and is subject to the whims of solar eruptions.

humans: survival at such low pressure would require the use of a pressurised spacesuit.

The Viking missions (1976) determined the other main components of the Martian atmosphere: *in situ* measurements confirmed the presence of the transparent and neutral gases nitrogen and argon. Traces of oxygen (O_2) exist, in negligible quantities compared with the oxygen content of the Earth's atmosphere, the result of photosynthetic activity in plants and bacteria. So, on Mars, oxygen derivatives, such as ozone (O_3), produced by photo-dissociation in the ultraviolet rays of the Sun, are rare. Such a small quantity of atmospheric ozone offers no protection from ultraviolet radiation: Mars has no 'ozone layer'. Therefore, this radiation can dissociate atmospheric CO_2 molecules into molecules of oxygen (O_2) and carbon monoxide (CO). Why Mars' carbon dioxide has not all been thus converted has long been a subject of debate: as far as is understood, the small quantity of water vapour in the atmosphere (see pages 149–151) provides the reactive agent OH, which triggers the rapid oxidisation of CO (recombining it with O_2) to form CO_2. This ensures that amounts of carbon monoxide and oxygen remain small compared with carbon dioxide (see table opposite). Mars Express and its SPICAM spectrometer, capable of measuring amounts of ozone and water vapour, have done much to increase our knowledge of this atmospheric chemistry. Another spectrometer, PFS, surprised researchers by detecting traces of methane (CH_4). This detection tested the instrument to its limits, but later (and no less difficult!) spectroscopic observations from Earth, it was claimed, confirmed this. Methane should be unstable in the Martian atmosphere, so its presence requires the existence of some sub-surface source: evidence for continuing volcanic activity? Or even of life, below ground?

3 Martian winds

Atmospheric circulation determined by the same laws as on Earth

Why are there winds on Mars (and on the Earth)? Two major mechanisms combine forces:

Thermal contrasts

Temperature differences created by solar radiation are the principal cause of atmospheric movements. Warm air expands and rises: so, for the same surface pressure, one finds more air at altitude, since warm air occupies a greater volume. Thermal contrasts thus lead to pressure differences at altitude, and masses of air at higher pressures (in warm regions) are 'pushed' towards low-pressure (cooler) regions. This movement we experience as winds.

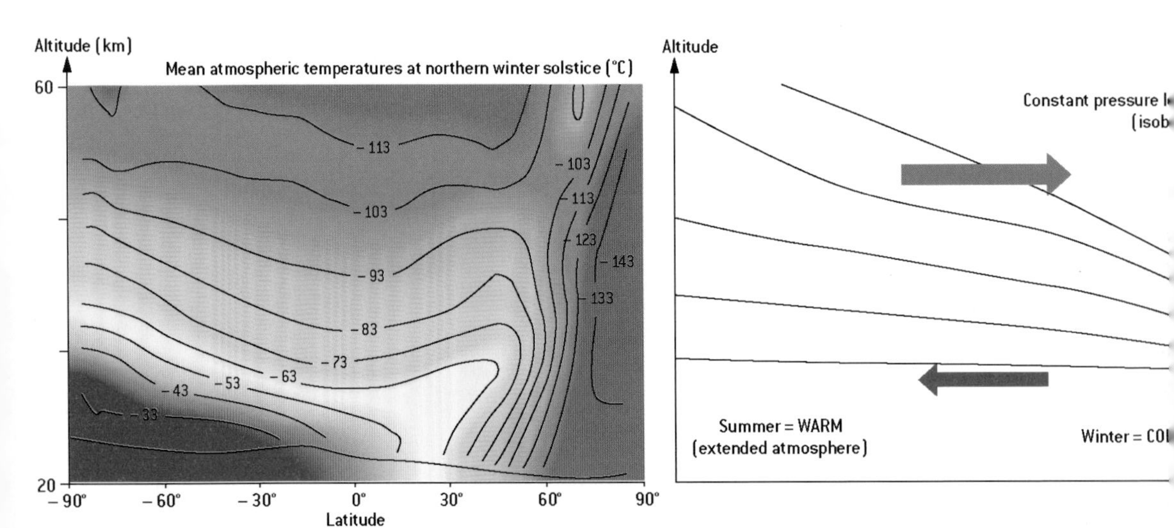

Wide thermal contrasts in the Martian atmosphere. The figure shows mean temperatures in the atmosphere at the winter solstice in the northern hemisphere, in degrees, as measured by the TES spectrometer on Mars Global Surveyor. The warm air of the summer hemisphere is less dense than the cold air of the winter hemisphere: at altitude, the 'summer air' is 'pushed' towards the winter hemisphere (Figure after M.D. Smith).

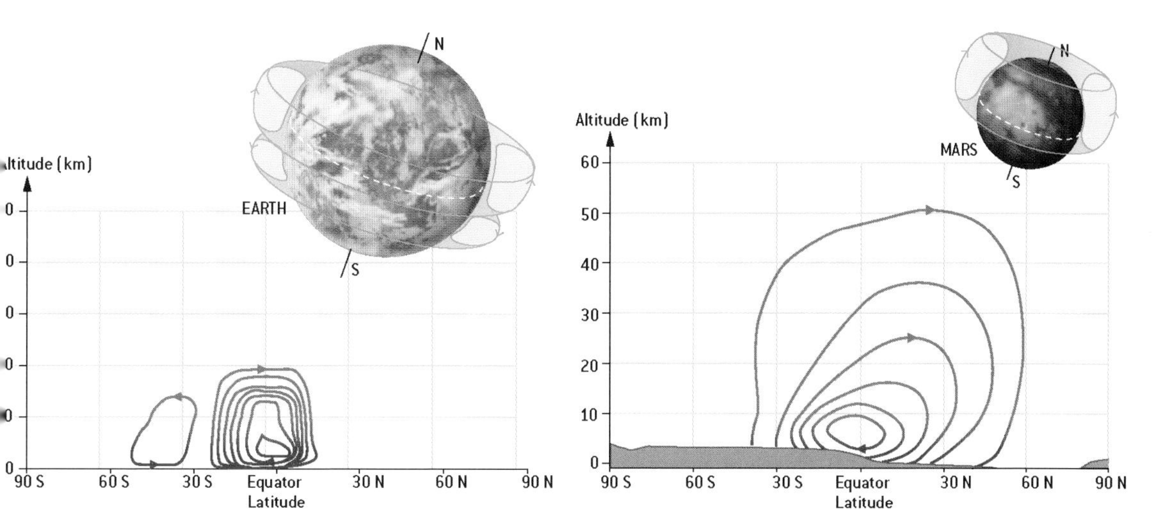

Meridional (north-south) circulation generated by thermal contrasts. At altitude, air masses in warmer regions are pushed towards cooler regions, where they cool and redescend. The air masses are then fed back towards warmer regions, which they encounter in the lower layers of the atmosphere. **On Earth**, this is known as the 'Hadley circulation', after the English physicist who first described the process in 1735. Around our planet, the atmosphere forms pairs of 'cells' between the intertropical zone and each of the hemispheres. **On Mars**, generally, only one large Hadley cell forms, involving both hemispheres and straddling the Equator. Also, the Martian cell is much more developed at altitude than its terrestrial counterparts. On Earth, the Hadley circulation is confined below the stratosphere, which prevents vertical movement of the circulation; on Mars, this cannot happen, as there is no stratosphere.

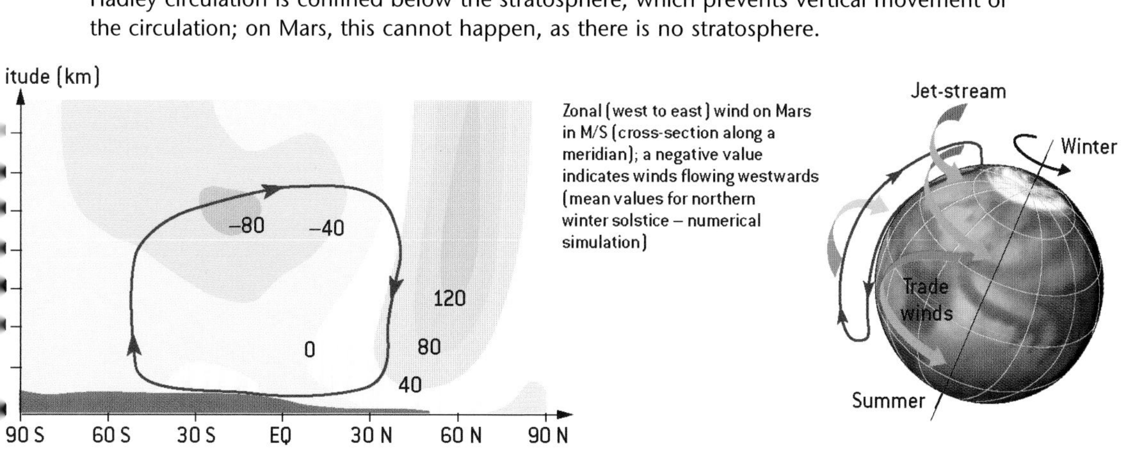

Zonal (west to east) wind on Mars in M/S (cross-section along a meridian); a negative value indicates winds flowing westwards (mean values for northern winter solstice – numerical simulation)

'inds in the Martian atmosphere, engendered by the meridional circulation and the rotation of the planet. On a tating planet, meridional (south-north) movements of air create strong lateral easterly or westerly winds, which ominate atmospheric circulation. Why is this? Near the surface, at the base of the ascending branch of the Hadley ll, the air follows the rotation of the planet on its axis. However, when air masses are drawn to the Equator, in the oper part of the cell, they move away from the axis of rotation. Much like an ice skater who extends her arms while inning, these particles will tend to turn less rapidly than the planet, creating 'retrograde' winds, blowing from st to west. As they draw nearer to the pole, the air masses also approach the axis of rotation. They will then turn ster than the planet, in the manner of the skater now drawing in her arms, to form a strong westerly jet around the ole, analogous to the terrestrial 'jet stream'. The same kind of process occurs at low altitudes (see pages 131–133).

The rotation of the planet

Air movements are also affected by planetary rotation (see figures on page 129). In fact, since the Earth and Mars have almost identical rotation periods and similar seasonal variations in insolation, the red planet exhibits meteorological elements not unlike those found on Earth. So, Mars has its westerly winds at around 50° N, a jet-stream (like the high-altitude current circling the Earth from west to east at mid-latitudes in both hemispheres), tropical trade winds, etc. At least, that is what we deduce from the scarce data available, provided by infrared detectors, and by the Viking landers. Computer models of the Martian atmosphere are used to help us analyse and interpret these data. Such studies also reveal fundamental differences between atmospheric circulation patterns on Earth and on Mars: the absence of oceans on Mars is a factor here. On Earth, the lower layers of the atmosphere are generally warmer at the tropics than at the poles, while on Mars, for most of the year, the thermal contrast is between the spring/summer (warm) hemisphere and the autumn/winter (cold) hemisphere. This phenomenon is due to the hyper-continental nature of the Martian climate, although, around the equinoxes, a more Earth-like situation pertains, and it is colder at the poles. In addition, the contrast in temperature between the hemispheres is heightened because of dust in the atmosphere, which absorbs solar radiation directly in regions where insolation is greatest.

The general mechanisms described above explain the global movements of the Martian atmosphere, as do the detailed illustrations on the previous pages.

4 Martian meteorology

Trade winds, thermal breezes and depressions at Mars' surface

For a specialist in the climate of planet Earth, a meteorological chart of Mars would not seem as 'extra-terrestrial' as one might imagine. The general patterns of dominant winds are similar on both planets. For example, at the latitude of Europe, winds blow mainly from the west, following a rhythm occasioned by the passage of high-pressure areas (anticyclones) and low-pressure areas (depressions) which succeed one another week by week in autumn and winter (see wind chart below).

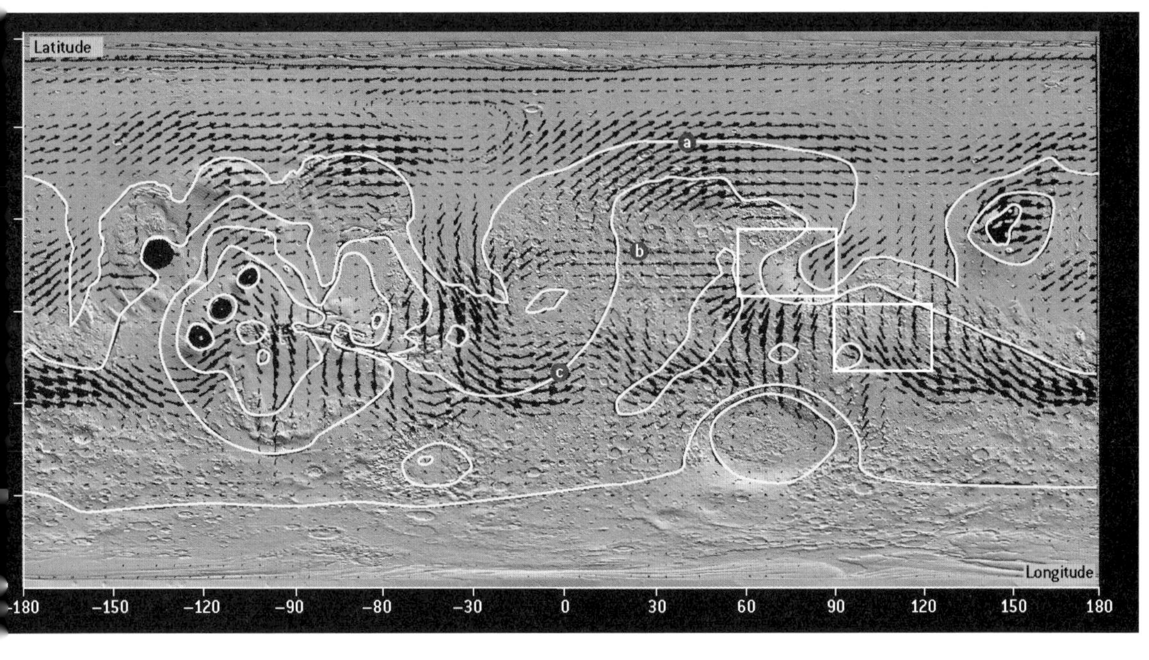

Chart of mean winds near the surface of Mars at the northern winter solstice, from a computer model generated at the Laboratoire de Météorologie Dynamique, Paris. At latitudes similar to those of Europe (35°-75° N), the Martian jet stream has the same effect at the surface as it does on Earth: the Martian winter winds tend to be westerly (a), especially on higher ground. Further south, the 'return' branch of the Hadley cell affects both northern and southern hemispheres in the lower layers. Diverted by the rotation of the planet, it creates easterly 'trade winds' to the north of the Equator between 0° and 30° (b), accelerates on crossing the Equator, and then causes strong westerly jets (similar to summer monsoon winds in India and central Asia) at around 30° S (c). (Courtesy MOLA Science Team.)

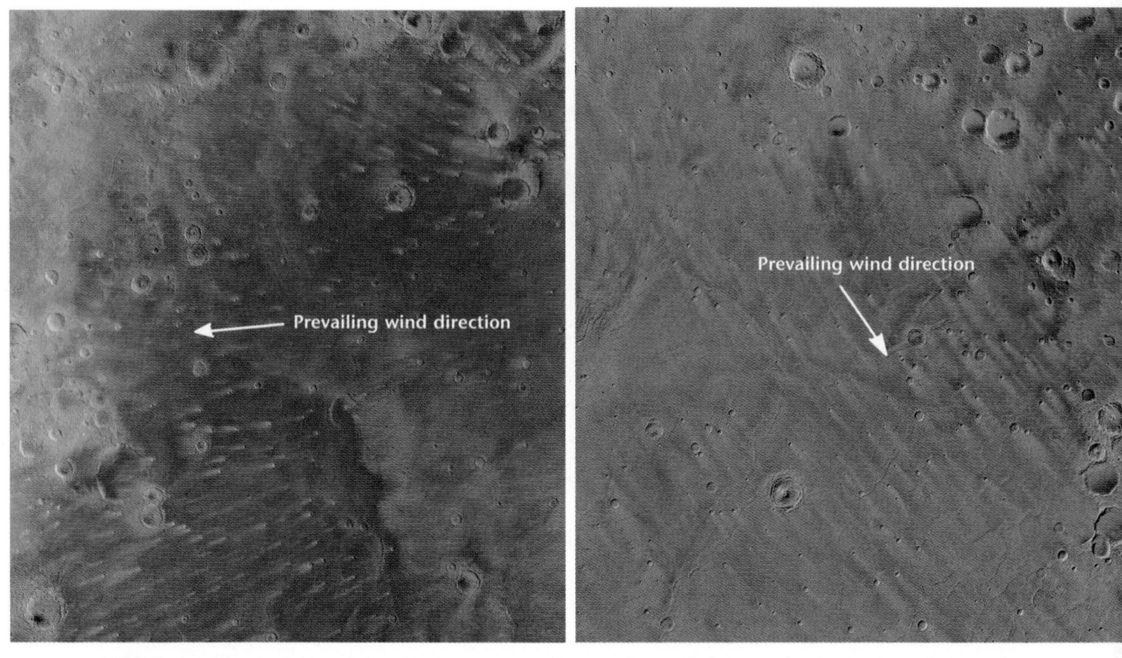

Wind streaks on the surface. Year after year, light-coloured dust settles in the lee of obstacles such as small craters, indicating the direction of the prevailing wind. These deposits are visible in areas where the ground is dark, as in these two areas near Isidis Planitia on either side of the Martian equator (shown as white rectangles on the wind chart). (Courtesy NASA.)

In Martian tropical latitudes, there is a regime of trade and monsoon winds rather like those on Earth where the climate is at its most 'continental', i.e. in Asia.

A closer look will reveal to the meteorologist that, as a result of the hyper-continental climate, the diurnal oscillation of winds, which is not a very important phenomenon on Earth, is reinforced on Mars because of the great differences between daytime and night-time temperatures. During the nights, the atmosphere near the Martian surface becomes extremely cold. This denser gas tends to flow, running down slopes and creating strong localised winds until dawn. In the afternoon, the opposite occurs, as warmed 'air' near the surface rises up slopes. These marked oscillations have been identified by all Mars probes.

It has been shown that diurnal oscillations of temperature and winds near the Martian surface have an indirect effect on all other layers of the atmosphere. Like any other physical medium, the tenuous Martian atmosphere vibrates, or, to be more precise, transmits diurnal frequency waves (one oscillation per day), or, in the manner of ocean tides, semi-diurnal frequency waves (two oscillations per day). This analogy gives them the name 'thermal tidal waves'. The diurnal oscillations, interacting with other winds, exert a considerable influence over atmospheric circulation on Mars – as well as presenting thorny theoretical problems to the would-be Mars meteorologist!

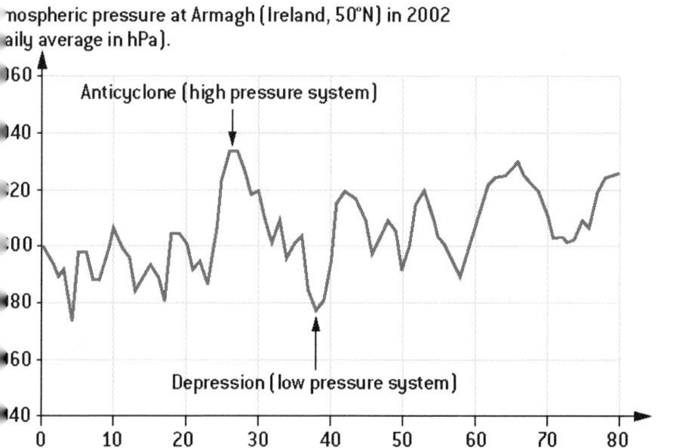

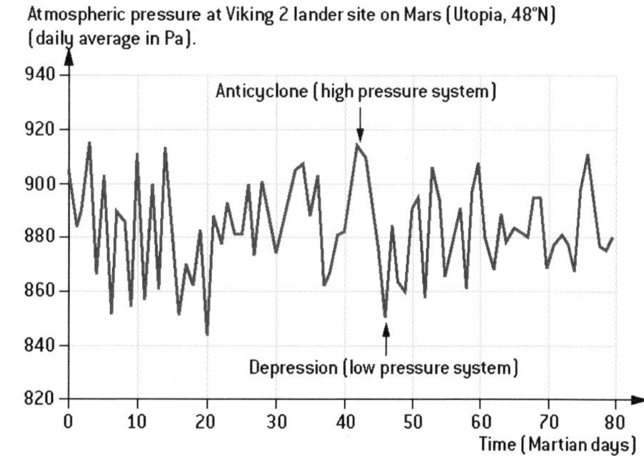

Low-pressure areas recorded on Earth and Mars at weather stations over an 80-day period in winter. Viking 2 measured variations in pressure on Mars similar to those recorded in Europe (compared with total pressures). Depressions sweep across mid-latitudes from west to east. They are produced by temperature differences between the tropics and the polar regions. Cold air tends to move beneath warmer air, creating instabilities and leading to planetary-scale waves in the atmosphere and oscillations in the path of the jet stream (see pages 128–130).

5 Dunes, ergs and erosion: footprints of the winds

Landscapes sculpted by the wind, reminiscent of Earthly deserts

The solid particles covering the Martian surface range in size from just a few micrometres in diameter (dust) to hundredths of micrometres (sand). The finest particles may be lifted and carried off by the wind, to become dust storms (see following pages). The largest grains may be blown along the ground, where they collide with other smaller dust grains, lifting them in turn off the ground (a process known as 'saltation'). As the years go by and the winds blow, sand and dust will accumulate to form dunes, which are particularly numerous on Mars.

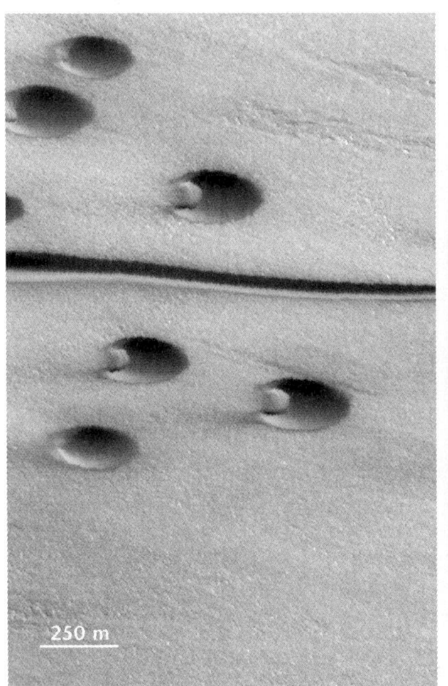

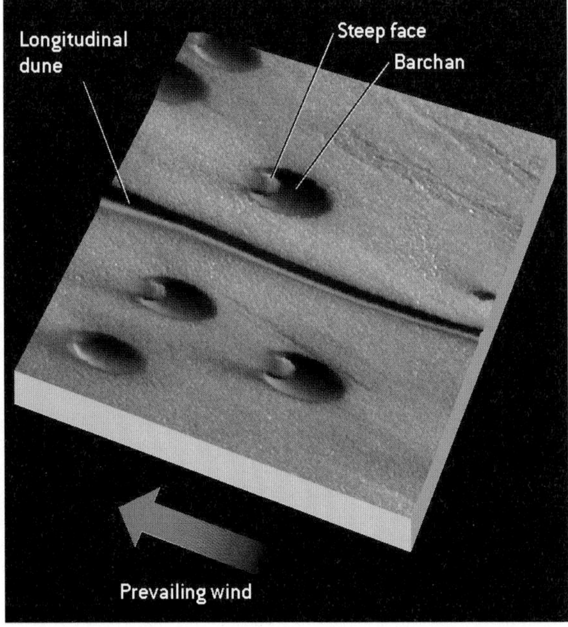

Linear dunes and barchans. Dunes on Mars may be linear in form, or crescent-shaped (barchans). In the case of barchans, grains accumulate on the leeward flank, and subsequent 'avalanches' cause that flank to be steeper. (Courtesy NASA, JPL, Malin Space Science Systems.)

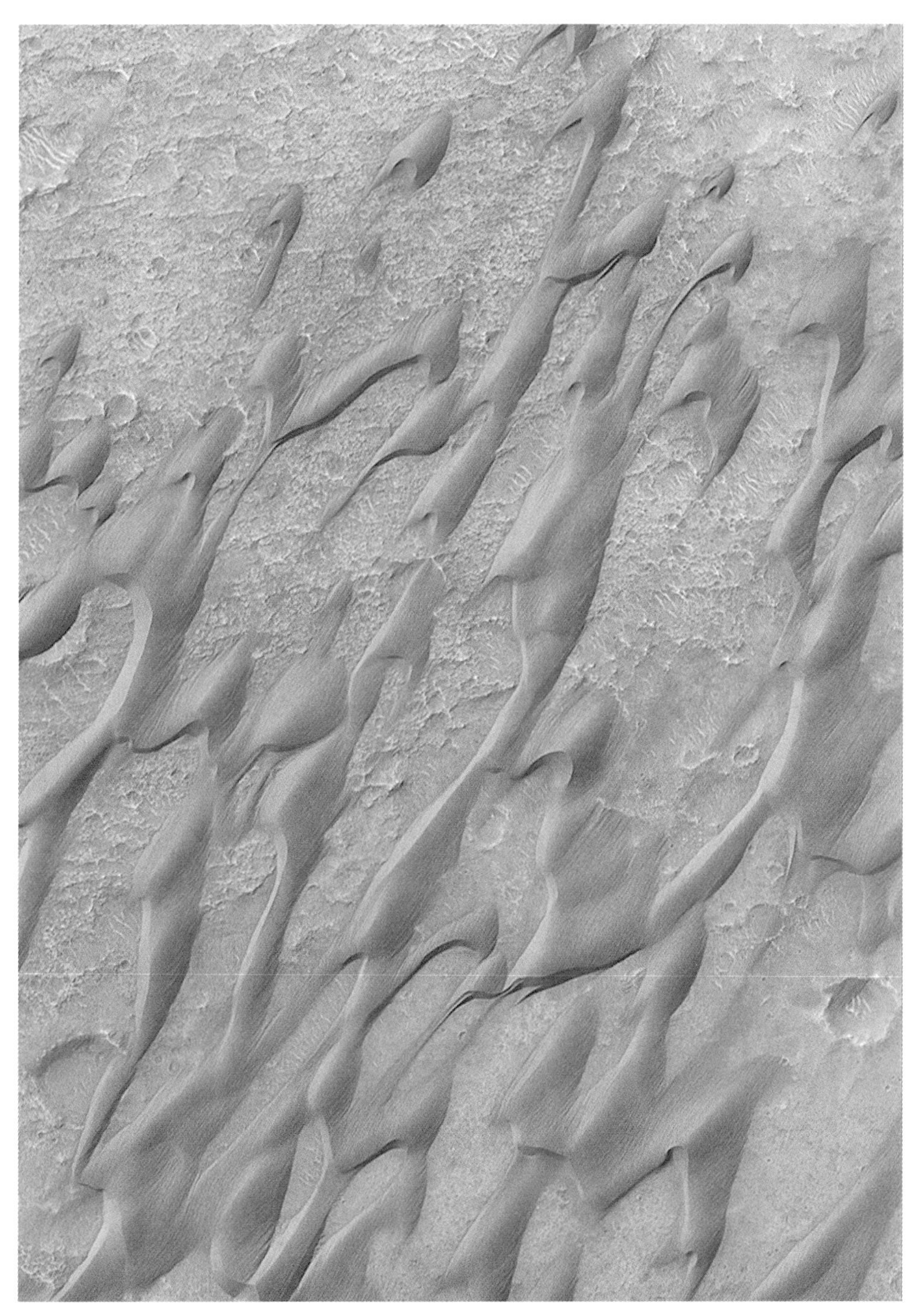

Dunes in Herschel Crater, south of the Elysium region. (Courtesy NASA, JPL, Malin Space Science Systems.)

Wind- and sand-abraded rocks. The left-hand image shows a triangular block, 40 cm long, near the Pathfinder landing site on Mars. Its shape and surface texture have been created by wind erosion. (Courtesy NASA, JPL.) This rock is similar to 'Dreikanter'-type (three-cornered) rocks on Earth, an example of which, from the Mojave Desert in California, is shown in the right-hand image. (Courtesy F. Costard.)

Dune fields, resembling the ergs of terrestrial deserts, are found almost everywhere along valley bottoms and in low-lying areas on Mars. Most of these dunes probably move as time elapses, though such motion has yet to be detected.

The transportation of grains by the wind, abrading the surfaces which they encounter, is also a factor in erosion. Images from Viking and Pathfinder clearly show that many Martian rocks have been eroded by wind-blown material, their surfaces exhibiting striations and centimetre-sized pits created by such abrasion. Some of these wind-eroded rocks have a characteristically triangular shape.

Wind erosion may not be a strong factor on Mars in the short term, given the tenuous nature of the atmosphere. However, over millions – or billions – of years, exposed rocks will be greatly affected by the shaping action of the wind.

Small dunes (false-colour image) photographed from the surface by the rover Opportunity, on the floor of Endurance Crater (see page 203). The tongues of sand in the foreground are no more than a metre high. The pink colour reveals that dust accumulates more on the crests than on the flanks of the dunes. (Courtesy NASA, JPL, Cornell.)

6 The dust cycle: ever-orange skies

An atmosphere laden with dust

The Martian sky over Chryse Planitia, photographed by Viking 1. On the ground can be seen a large boulder, named 'Big Joe' by the Viking team. It is more than a metre high and is covered by wind-blown dust. (Courtesy NASA.)

The first colour picture to be transmitted from the surface of Mars arrived on Earth in July 1976. The digital image, from Viking 1, was immediately processed and colour-balanced by NASA technicians. Without any previous reference by which to work, they 'pushed' the colour adjustments to arrive at a somewhat

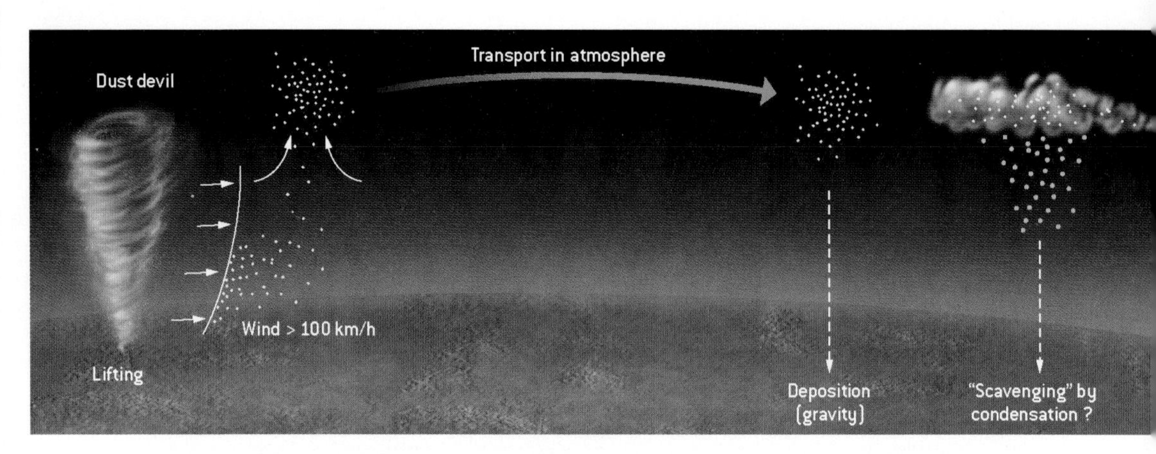

The dust cycle in the Martian atmosphere. The three main phases are lifting, transport by the atmosphere and deposition. A fourth phase of 'scavenging' by clouds and precipitation may play a small rôle on Mars – it is an essential phase in the dust cycle on Earth, being the reason why Earth's atmosphere stays relatively clear in spite of its many dust storms.

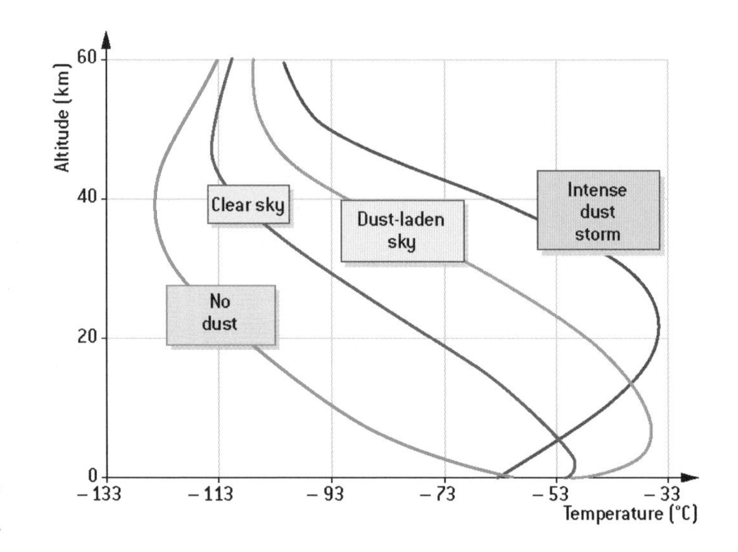

Dust: ruler of the Martian climate. Variations in mean temperature according to an atmospheric model simulating effects of dust above Viking 1.

Earth-like grey-blue sky, but later and better calibration brought out the real (and unexpected) hue of the Martian sky: not blue, but russet, orange, salmon-pink... The presence of mineral dust particles of the order of one micrometre in diameter, suspended in the atmosphere, creates this effect.

How can such a thin atmosphere, with a density so low that the strongest winds would be imperceptible to future Mars astronauts, lift and hold aloft so much dust? There is as yet no exact answer to this question. Wind-tunnel experiments, simulating conditions on Mars, have shown that occasional winds

Deposition of dust on Spirit's calibration target. After 416 sols (Martian days), Spirit had become covered with a fine layer of dust, which considerably reduced the efficiency of the solar panels in supplying current. Fortunately, a 'dust devil' (see pages 143–145) swept clean the rover and its solar panels during sol 420. Then, Spirit was able to perform as well as it had done in its first days. (Courtesy NASA, JPL, Cornell.)

of 100 km/h are able to move particles of the order of 100 micrometres, which, in their turn, dislodge, by saltation (see page 134), dust fine enough to remain aloft for a considerable period before settling again (sedimentation). In Earth's atmosphere, quantities of dust are not normally seen, as rain and snow regularly 'scavenge' it out. Such a phenomenon could exist on Mars, but it would be far less efficient.

Dust in suspension causes great fluctuations in the opacity of the atmosphere, leading to considerable variations in average 'air' temperature. Even when the Martian atmosphere is relatively clear, the temperature at an altitude of 20 km is 30°C higher than it would be in the total absence of dust. When dust storms are in progress, this value may attain 80°C! On Earth, such variability is unknown. During the day, Martian dust absorbs radiation from the Sun, warming the high atmosphere and cooling the surface. At night, however, infrared emission from the dust warms the surface. Therefore, the presence of dust tends to diminish the contrast between daytime and night-time surface temperatures (see pages 122–124); whereas mean surface temperatures are not greatly modified by atmospheric dust throughout the day-night cycle.

7 The dust cycle: planet of storms

Every day, many dust storms move across Mars

Most of the dust in the Martian atmosphere is raised by countless localised storms, everyday events on Mars and observed by space probes. Most of these events occur in association with atmospheric low-pressure areas in high and mid-latitudes, or around the edges of the seasonally changing ice caps (see pages 155–157), where temperature contrasts between the carbon dioxide 'snow' and the

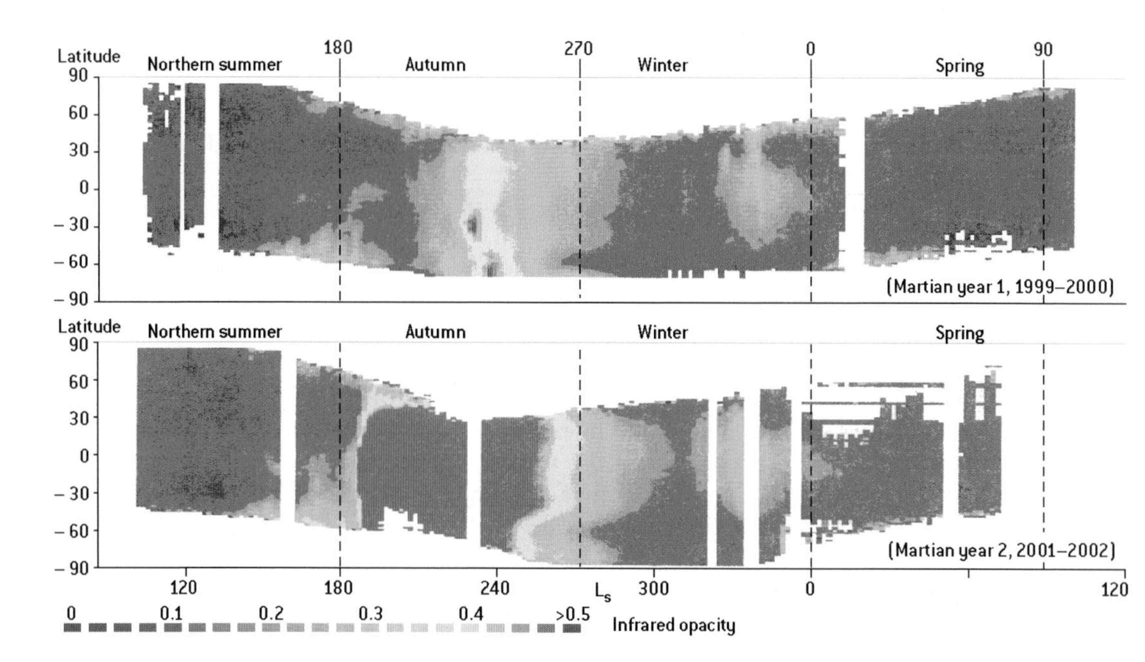

Variation in amounts of dust in the Martian atmosphere as a function of latitude, as measured by the TES spectrometer aboard Mars Global Surveyor. Very clear atmosphere appears in violet, and heavily dust-laden atmosphere in red. Events in the 'season of great storms' vary considerably from year to year. For example, in 1999, only a few regional dust storms were observed, yet in 2001, a planet-wide storm lasted for months. The figures clearly show the lifting of dust in localised storms around the rim of the seasonal polar caps (the caps are shown in white here, since infrared measurements of dust were not possible at low temperature). (Figures after M.D. Smith).

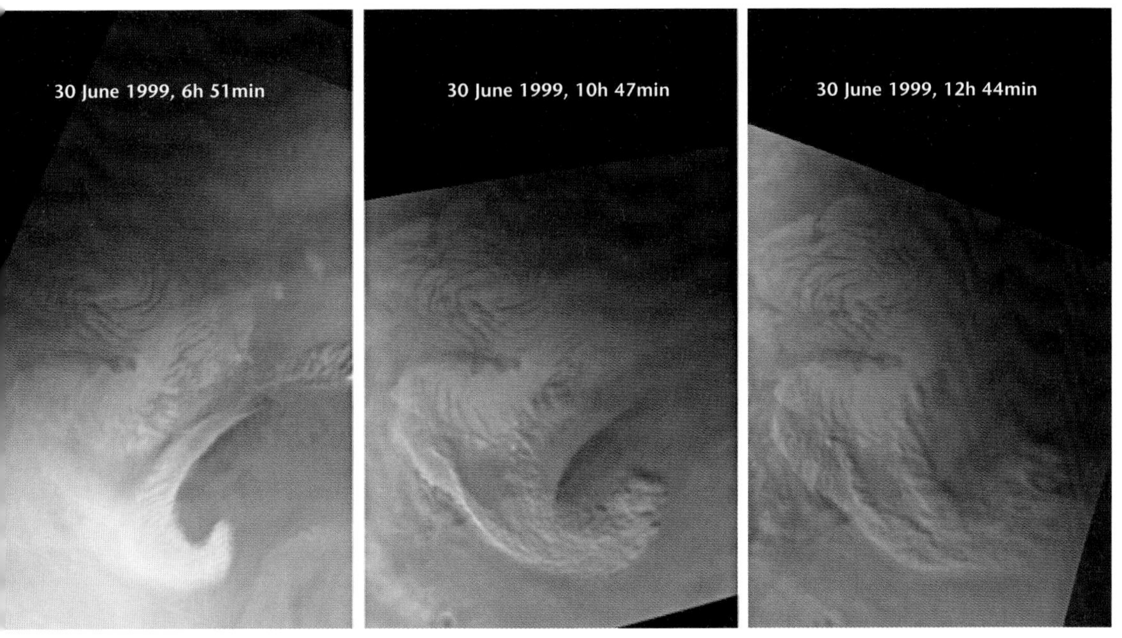

Development of a dust storm in the north polar region in summer, hour by hour (Mars Global Surveyor). (Courtesy NASA, JPL, Malin Space Science Systems.)

ice-free surface engender strong thermal breezes. Every year, dozens of Martian storms attain regional importance, extending across thousands of kilometres. They may last for several days, especially in southern spring and summer. Observations of atmospheric dust tell us that the year can be divided into two distinct periods: a 'clear season' during northern spring and summer, and a 'dust season', when great storms occur, during autumn and winter. In the clear season, storms are relatively few in number, and there is not much dust in suspension in the atmosphere. One clear season very much resembles the next: typical small storms arise on similar dates, and the usual quantities of suspended dust are observed. However, the major meteorological events that characterise the dust season vary from one year to the next. In some years, there may be one or more truly global storms, shrouding the Martian surface for months on end (see pages 146–148).

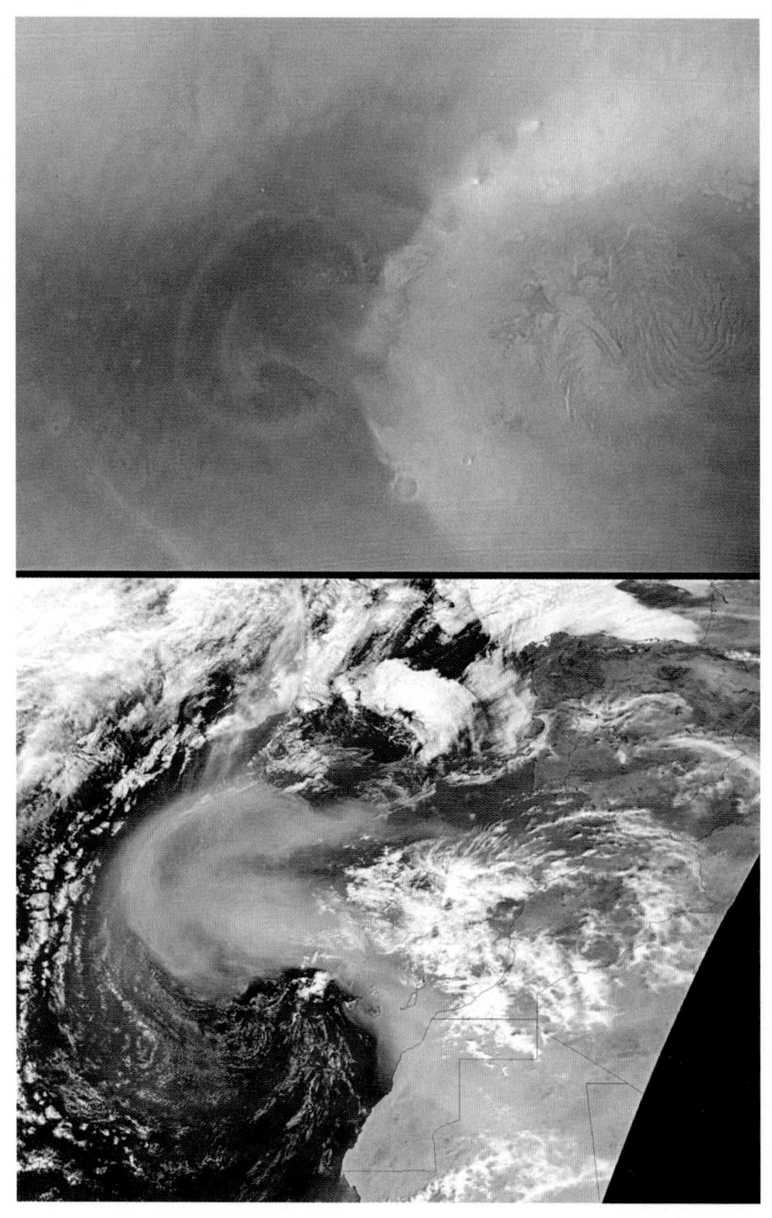

A typical example of a dust storm on Mars (upper image), raised by violent winds blowing around the edge of the seasonal polar carbon-dioxide cap (see pages 155–157). This storm was photographed by Mars Global Surveyor, near the north polar cap in August 2000. (Courtesy NASA, JPL, Malin Space Science Systems.) Around 900 kilometres across, it bears an interesting resemblance to a Saharan dust storm which occurred a few months earlier on Earth (lower image). (Courtesy NASA, GSFC and Orbimage, SeaWIFS Project.) The structure in both cases is the same: to the rear, a central 'jet' fed by the storm, whose zone of activity is limited, and to the fore, a mushroom-shaped front of spreading dust.

8 The dust cycle: whirlwinds of dust

In the heat of the day, dust devils roam the Martian plains

Observations from Viking in the 1970s, and from later probes such as Pathfinder, Mars Global Surveyor and Mars Exploration Rover, revealed that atmospheric dust on Mars is also lifted in the form of vortices commonly known as 'dust devils'. By studying their shadows on the ground, which are far easier to see than the dust columns themselves, researchers have estimated that these phenomena may be of much greater dimensions then their terrestrial counterparts. Martian dust devils may reach 10 kilometres in height, and be some tens of metres, or even hundreds of metres, in diameter. They are phenomena of the Martian afternoon, when temperatures at the surface are at their highest, and atmospheric convection is therefore at its most intense. A dust devil draws in surrounding 'air' and concentrates it at its centre. According to the same principle as that which makes ice skaters rotate faster as they pull their arms inwards, the masses of air sucked in towards the dust devil's axis of rotation are strongly accelerated, resulting in violent

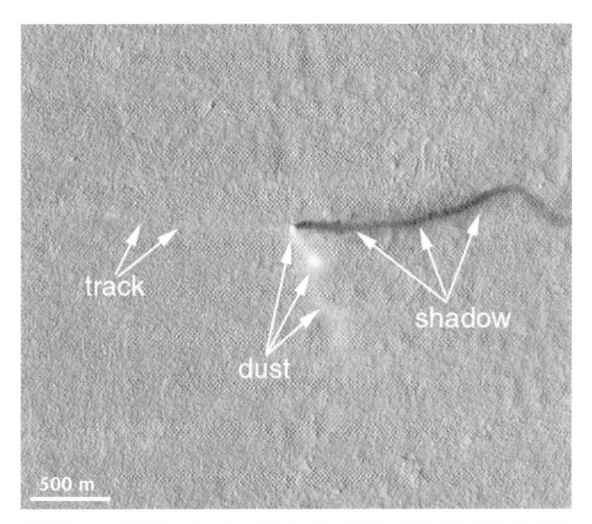

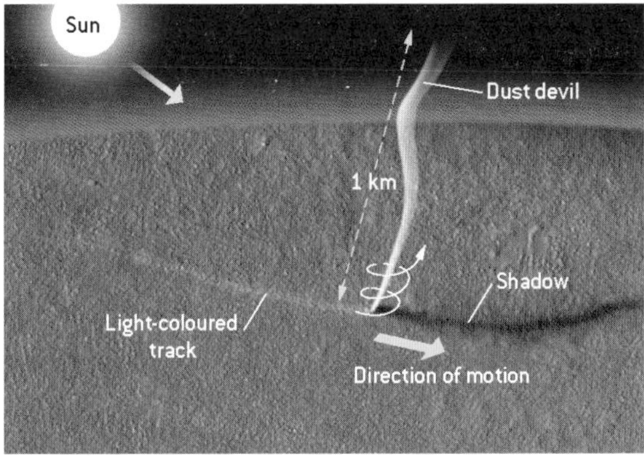

A dust devil in action, photographed by Mars Global Surveyor at about 14.00 local time. (Courtesy NASA, JPL, Malin Space Science Systems.)

transverse winds. In combination with the vertical aspiration at the centre of the dust devil, these winds succeed in lifting large quantities of dust. So these dust devils contribute to the supply of dust in the Martian atmosphere, though in what proportion remains to be discovered. One highly visible feature of the planet's surface are the innumerable 'graffiti' scored across it for great distances by dust devils (see images below and on facing page). These tracks are so easily seen because the ground 'cleaned' by a dust devil normally looks darker than the fine layer of dust all around.

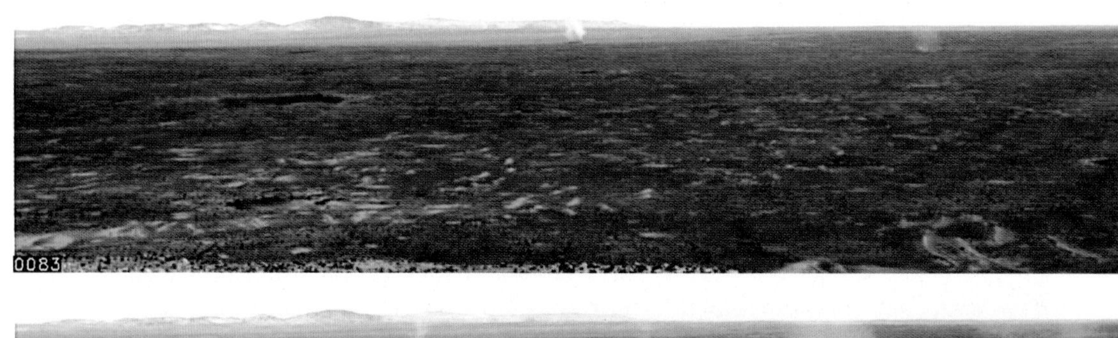

Dust devils in Gusev Crater, observed from the top of a hill by the rover Spirit on 13 July 2005. The two lower images are separated in time by 34 seconds, the first two images by 107 seconds. (Courtesy NASA, JPL, Texas A&M.)

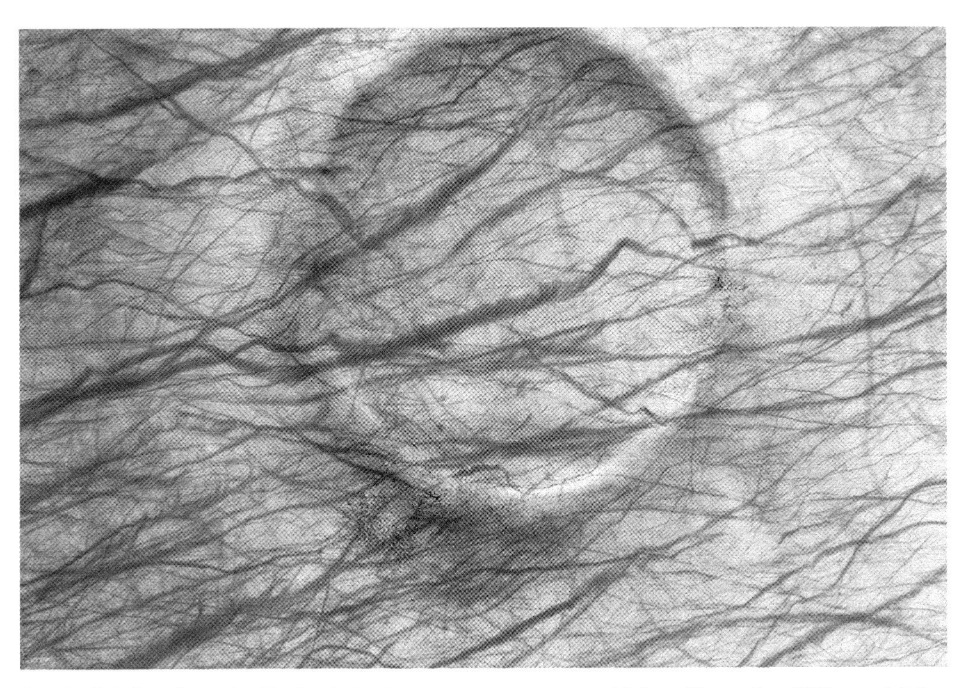

Tracks left by dust devils. Such tracks cover many areas of Mars. These 'graffiti' are visible even at quite high latitudes, for example here, at 57° S, around an ancient impact crater. (Courtesy NASA, JPL, Malin Space Science Systems.)

9 The dust cycle: global storms

When the entire planet disappears beneath dust

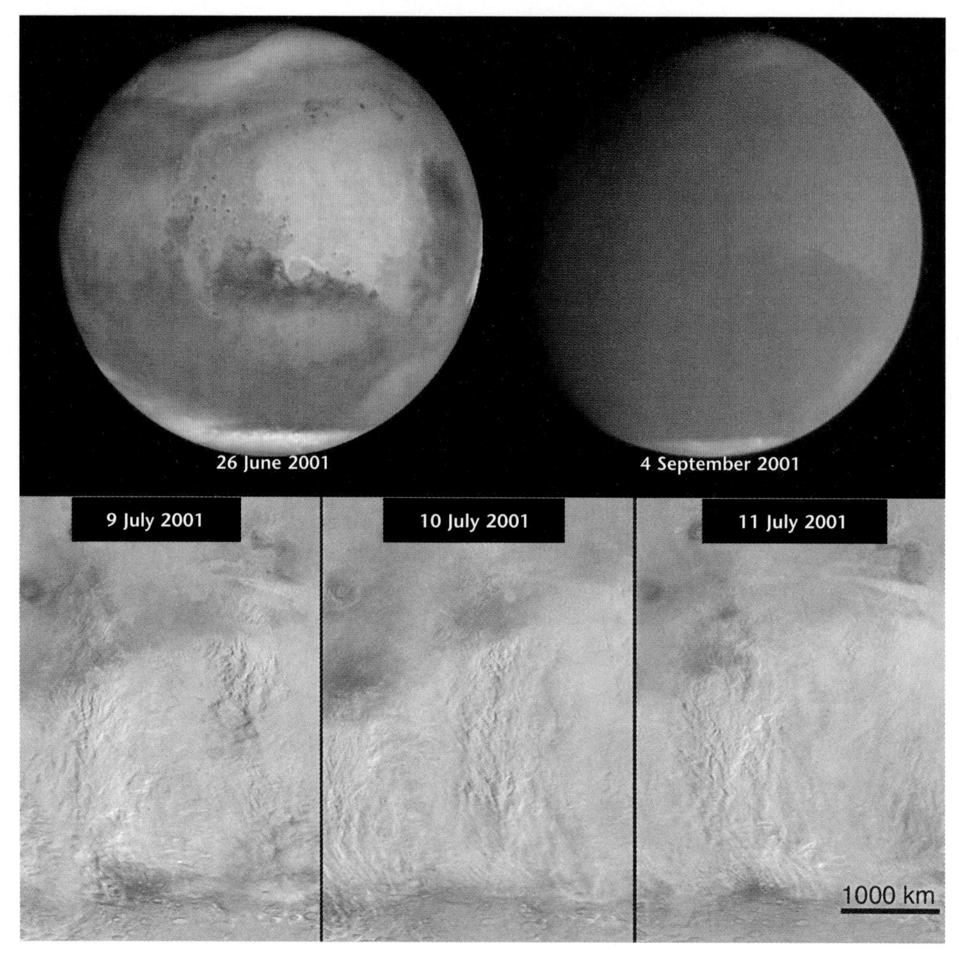

The dust storm of summer 2001. During the last days of June 2001, a gigantic dust storm brewed up on Mars. Originating near the southern polar cap and moving towards Hellas, it travelled some 4000 km in two days, crossing the Equator and triggering upsurges of dust in many regions. The planet disappeared beneath the dusty veil until the end of September, as the two upper Hubble Space Telescope images show. Images courtesy NASA STScI. Later analysis of the observations revealed that this was not truly a global event, but rather a chain of more local events, with dust being raised actively in certain regions (e.g. south of the Tharsis volcanoes on 9, 10 and 11 July, in the lower images). (Courtesy NASA, JPL, Malin Space Science Systems.)

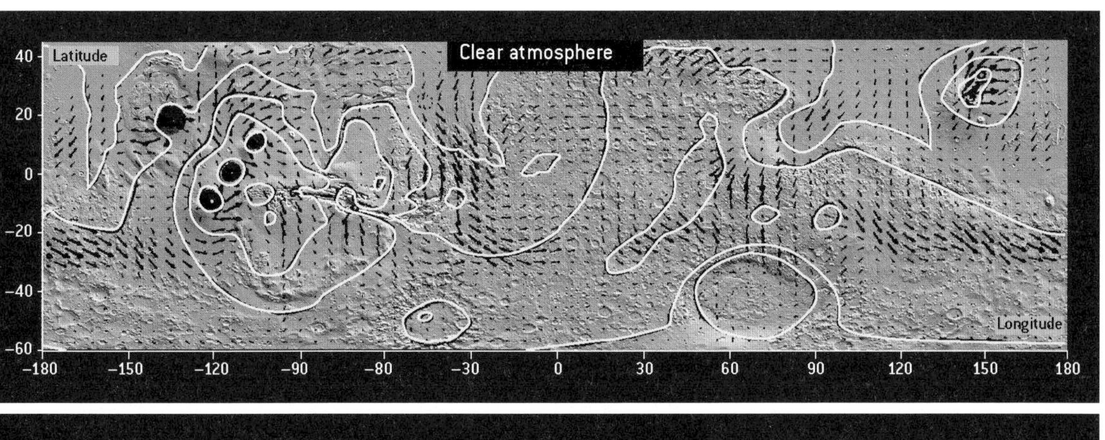

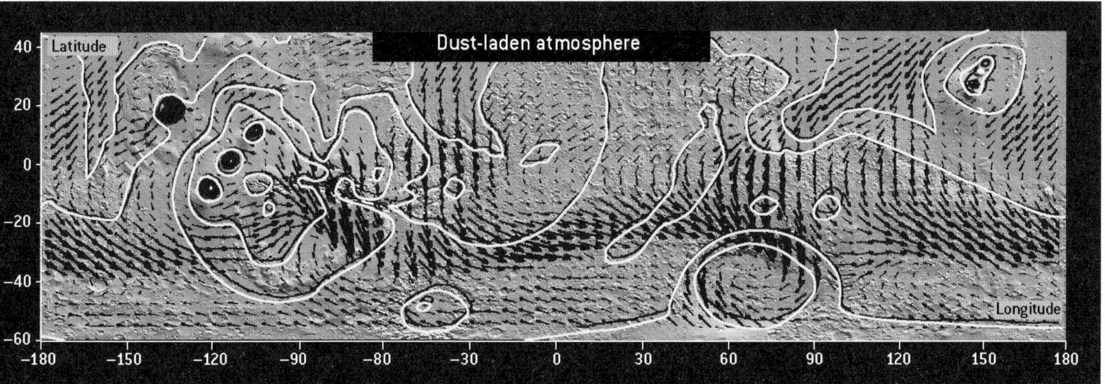

How does a local dust storm become a planet-wide event? The mechanism is as yet poorly understood. What we do know is that the lifting of dust into the atmosphere causes considerable meteorological changes. By absorbing the Sun's rays, the dust warms the atmosphere locally, creating or increasing temperature differences, the driving force behind winds and atmospheric circulation. On the global scale, a little dust lifted into the tropical atmosphere of Mars' southern hemisphere (where many local dust storms occur during summer) intensifies the circulation within the Hadley cell and thus the 'trade winds' (see pages 128–133). Models show that this strengthens the winds especially at the southern tropics. These winds pick up more dust, and the dust causes stronger winds. . . the machine fuels itself. This kind of mechanism also operates on a local scale. Global storms are probably triggered by a combination of local and global phenomena. In the light of these facts, why do such storms occur in some years and not in others? The answer is still elusive. (Figures inspired by F. Hourdin.) (Courtesy MOLA Science Team.)

During each southern spring, when Mars is closest to the Sun and insolation is at its peak, large numbers of local (and sometimes regional) dust storms occur in the southern hemisphere. In some years, one or two of these will grow into an exceptional meteorological event, unique in the solar system: a planet-wide dust storm. For many months, almost the whole of Mars will be veiled in dust.

The first of these events to be observed was that of 1956, when Mars was closer to the Earth than usual. Astronomers were surprised by the extent of the phenomenon, and disappointed at being unable to see anything through it. The great American populariser and champion of astronomy Carl Sagan wrote that, at the time, 'Mars was about as interesting as a baseball with no seams'. The phenomenon was recorded again in 1971, as the American probe Mariner 9, and then two Russian probes, Mars 2 and Mars 3, went into orbit around the planet. They were witnesses to the most intense global dust event ever seen. As the dust began to settle, Mariner 9 was still operating, and managed to secure photos of the surface (see pages 187–189). During the next Martian year (1973 on Earth), another planet-wide storm occurred; and no less than two, a hundred days apart, were unleashed during the first year of Viking's 1977 mission. Nothing of the sort happened during the next two Martian years, but in Viking's fourth year (1982) yet another global dust storm was recorded. Nearly 20 years went by before the next event, when Mars Global Surveyor told us of a new global storm. The 1970s saw the greatest concentration of such storms, as recorded between 1956 and the present. Why should that decade be so exceptional? The answer may have more to do with the way in which astronomers observe the planet, and with the space missions themselves, than with any climatic anomaly. As it happens, 1971 and 1973 were years of better visibility during the storm season, as was 1956: resolution was good. The geometry of the two planets' orbits lends itself to more favourable observations at intervals of about 16 years.

10 The water cycle: vapour, frost and ice – water on Mars

The atmosphere of Mars transports water as vapour and clouds

Mars may no longer have oceans, or even 'canals', but there is still plenty of water locked into the planet, even to the extent of its displaying a seasonal hydrological cycle.

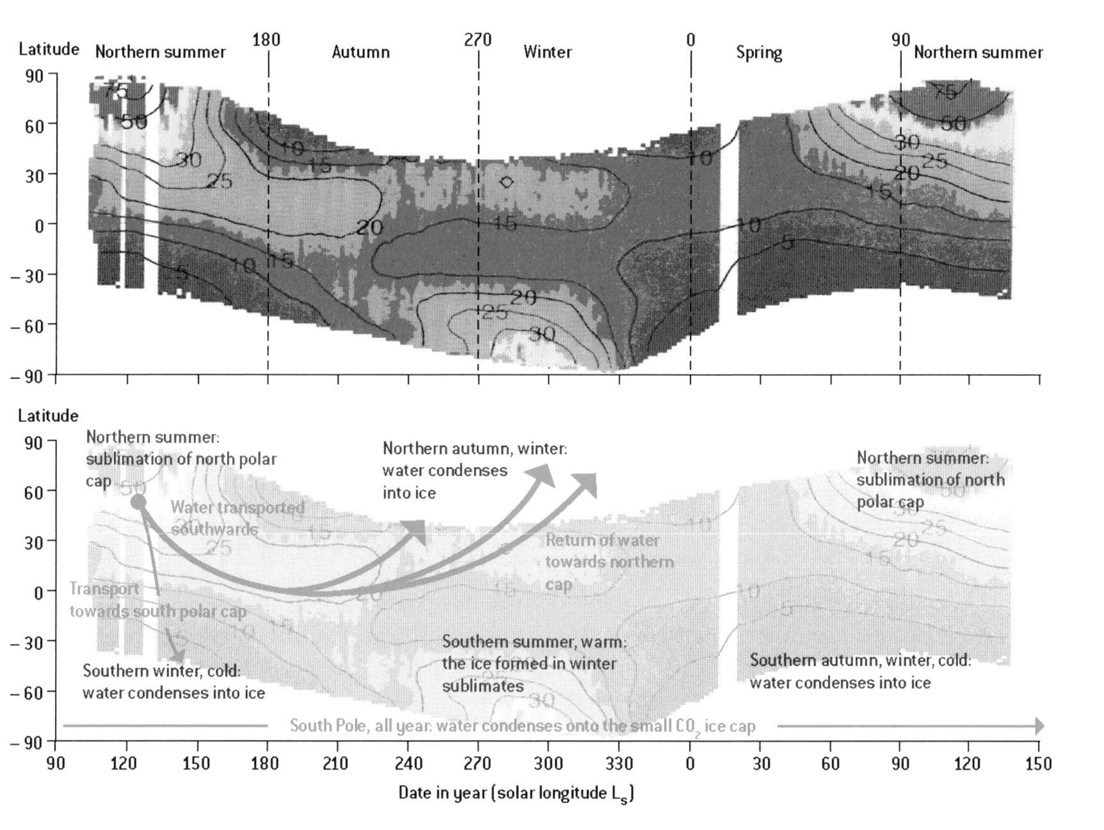

The annual water vapour cycle observed by the TES spectrometer on Mars Global Surveyor. The quantity of water vapour in the Martian atmosphere is represented here as a function of season and latitude, and expressed in 'precipitable micrometres' (i.e. the depth of liquid water formed if all atmospheric water condensed onto the surface). (After M.D. Smith).

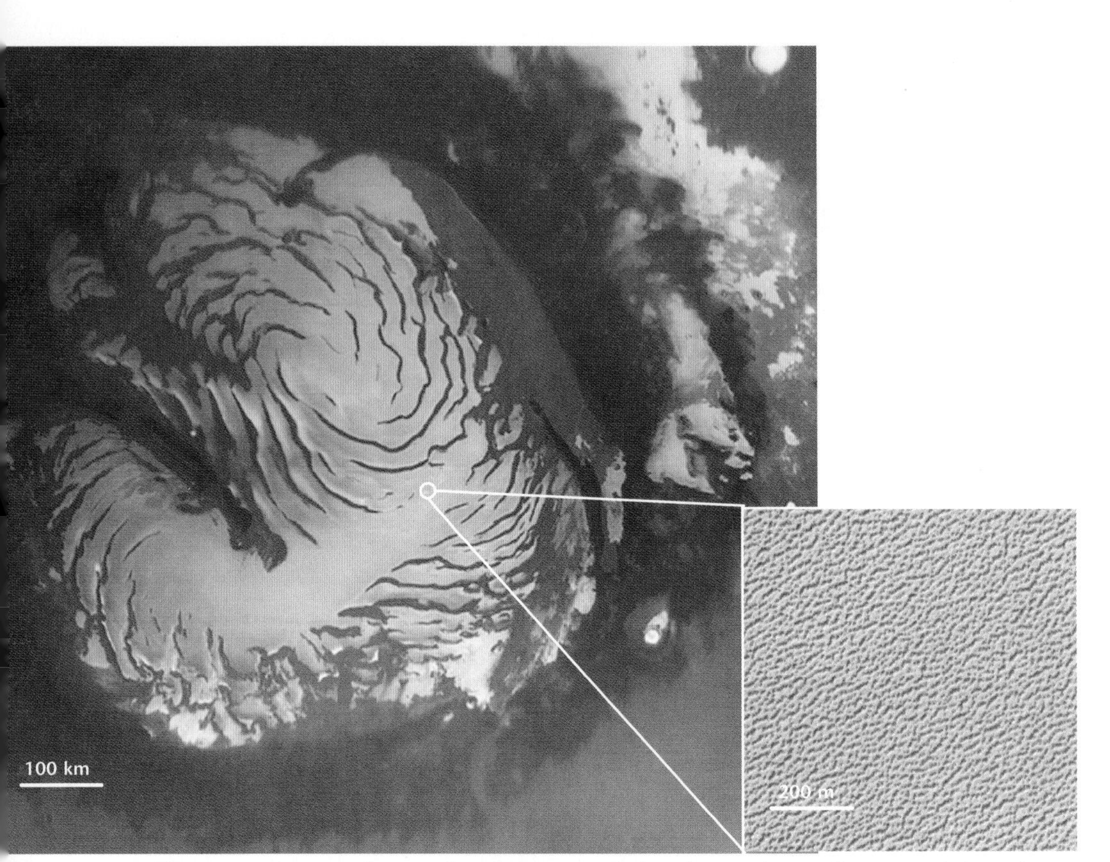

The permanent northern polar cap: a vast glacier of water ice, sediments and dust, riven with curving troughs. The cap, more than 1000 km across, is covered with a layer of relatively pure, white water-ice interacting with the atmosphere. Warming in summer makes it a source of atmospheric water vapour. These exchanges have made its surface a sculpture of pits, bump and knobs about twenty metres wide and two metres deep (see close-up at right). (Courtesy NASA JPL, Malin Space Science Systems.)

Everything starts at the north pole. As time has passed, an immense glacier of water, dust and sediments has formed. In summer, the Sun warms its icy surface. The ice does not melt, as the pressure on Mars is too low to allow water to exist in the liquid state: instead, it sublimates directly into the 'air'. So, every summer for a few months, the north polar region becomes a source of water vapour which is transported in the atmospheric circulation towards other latitudes.

The amount of water involved is not great: if all the water in the Martian atmosphere fell to the surface, there would be enough to form only a few tens of micrometres of ice in the 'wettest' regions. In spite of the comparative lack of water in the atmosphere, saturation is often reached at such low temperatures: clouds form (see pages 152–154), and frost can appear at the surface. Water

Frost on Mars. In 1978, the Viking lander observed frosts covering the Utopia plain from the end of autumn into the winter. For years, scientists debated the nature of this frost: is it water ice, or carbon dioxide produced by condensation in the atmosphere? Nowadays, the consensus among researchers is that it is water ice, so the photo echoes scenes familiar to us on Earth. (Courtesy NASA.)

vapour can also diffuse into the porous regolith which forms the uppermost layer of the planet. This set of processes represents Mars' water cycle. The essential question arising here is whether the cycle is a closed one: does the water periodically return to its initial state, as ice in the north polar cap? And the reply is 'Yes, but...', for there exists on Mars an area where water condenses but never leaves: the south pole. This is due to the presence of a layer of carbon dioxide ice permanently covering the southern polar cap (see pages 164–166). The surface here is at such low temperatures (–130°C) that any water present will exist only as ice. The ice cap locks water in a veritable deep-freeze. So the long-term water cycle on Mars seems to be a one-way journey, with water being transported from the north pole to the south pole. This dynamic cannot have been a feature of Mars throughout its history, or all water would have been removed from the north pole. It is very probable that, at some time in the past, the cycle operated in the opposite direction.

11 The water cycle: clouds and fogs

Thin clouds without rain or snow

Clouds around Olympus Mons. The most spectacular Martian clouds are those which form near high ground, and especially around giant volcanoes such as the Tharsis groups, Olympus and Elysium, in particular during northern summer when the atmosphere is most heavily laden with water vapour. On encountering a mountain, moving air masses are forced upwards and become considerably cooler. Water vapour condenses. As on the Earth, the atmosphere assumes a wave-like motion in the lee of the relief, and clouds form at the crests of the waves (Viking image with colour rendered by artist). (Courtesy NASA.)

On Earth, clouds play an essential part in meteorology and climate. On the one hand, they release heat as they condense (i.e. when water vapour turns into clouds), and represent a great source of energy in our dynamic atmosphere; on the other hand, their radiative properties (their capacity to reflect or trap radiation) greatly affect the energy balance of this planet.

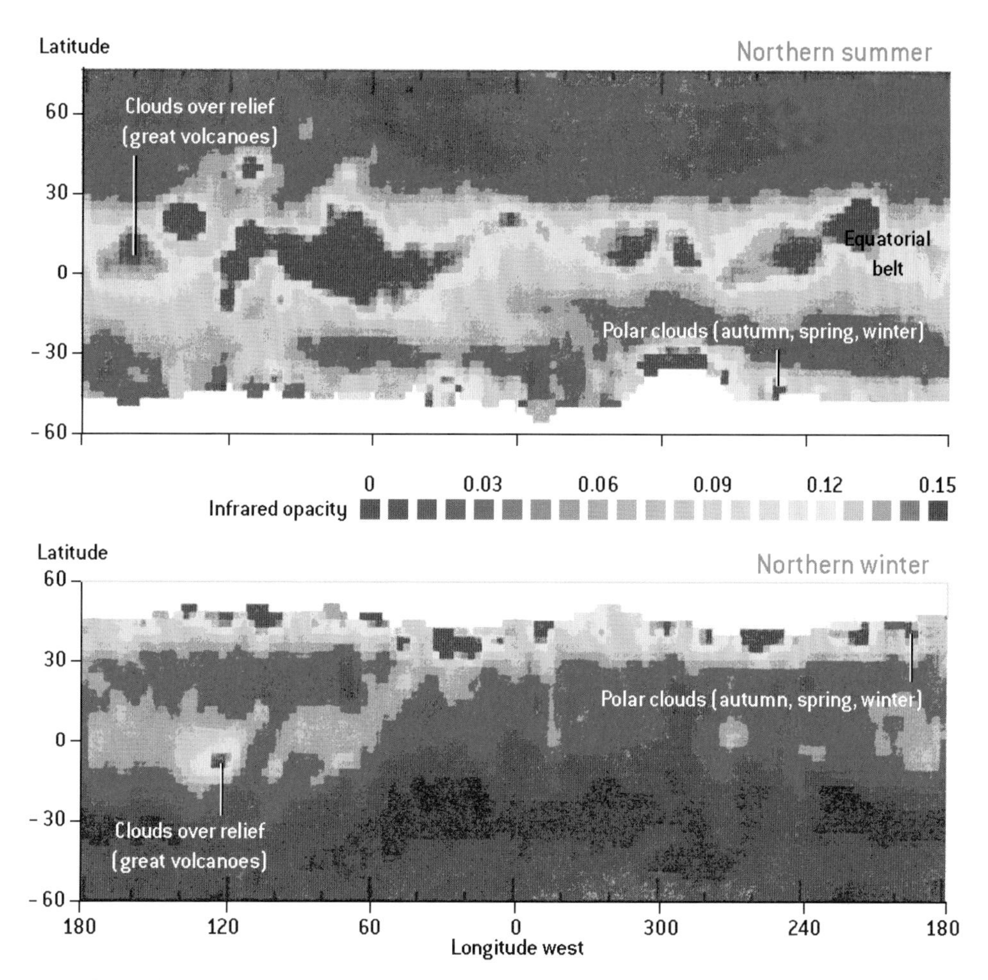

Two charts showing the infrared opacity of clouds measured for two opposing seasons by the TES spectrometer on Mars Global Surveyor (Figures after M.D. Smith).

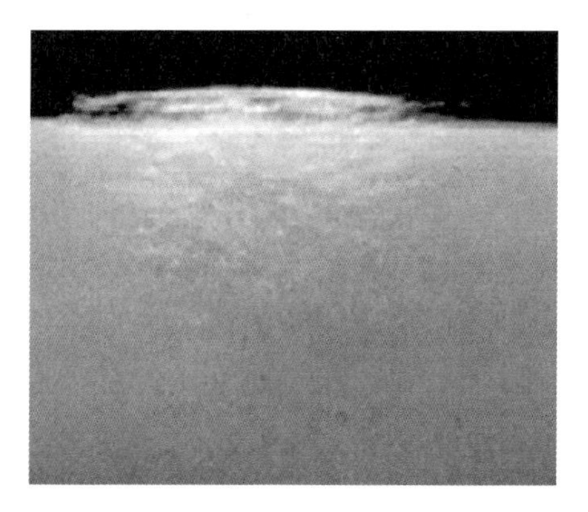

The equatorial cloud belt, seen in section by Mars Global Surveyor. In spring and summer (northern hemisphere), equatorial clouds may rise to an altitude of 30 kilometres. (Courtesy NASA, JPL, Malin Space Science Systems.)

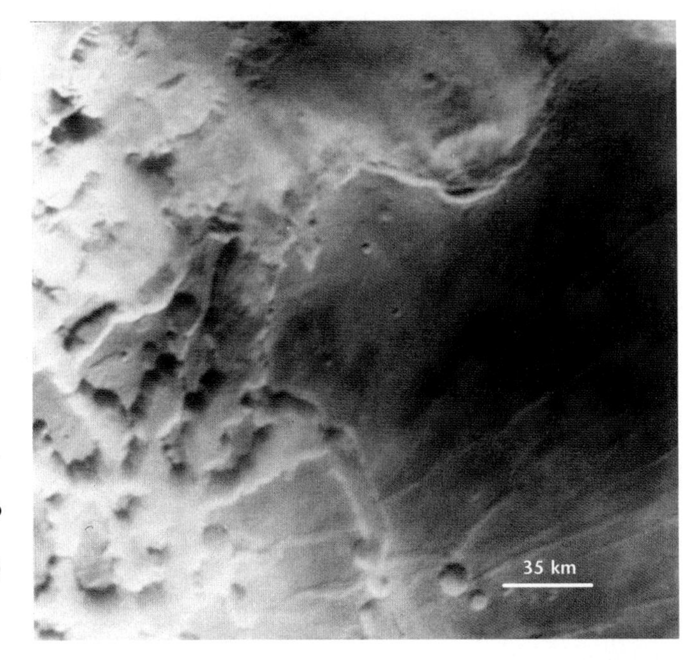

Morning mist in a network of valleys to the west of Valles Marineris. Nocturnal cooling sometimes triggers the formation of mists close to the surface. Observations by the Viking and Pathfinder landers indicate that these mists occur every night at numerous locations on Mars, generally evaporating as soon as the Sun appears. It is a common sight on Earth for morning fogs to linger in certain deep valleys. (Courtesy NASA.)

On Mars, it is a different story. Temperatures, pressures and amounts of water involved are considerably lower. However, Martian clouds are still of interest to planetologists. There are various types of clouds, the 'icing on the cake' of Martian weather. We know, for example, that the fine ice crystals (1 to 5 micrometres in size) which constitute the clouds will sublime before ever reaching the ground: their velocity of descent and the rate at which they aggregate are too low for rain ever to fall on Mars.

Where do we find clouds on Mars?

Almost every year, from the end of spring and through the summer in Mars' northern hemisphere, a band of clouds, visible to observers on Earth, girdles the Martian equator. The origin of this cloudy manifestation lies at the north polar ice cap, where vast quantities of water vapour are released into the atmosphere, which is cold because little dust is present. Transported to tropical latitudes, this water vapour is rapidly borne aloft in the ascending branch of the Hadley cell (see pages 128–130). As it rises, it encounters ever colder atmospheric layers, until it reaches a level at the limit of saturation, where it condenses. The Martian cloud belt is the equivalent of the intertropical cloud belt on Earth, where its periodic occurrence gives rise to the 'rainy seasons'.

Now, when autumn and winter come to the polar regions, there is marked atmospheric cooling. The mass of air moving towards the poles condenses, and many different cloudforms are seen, some diffuse, and others highly structured. This is what astronomers refer to as the 'polar hood'. It is especially visible in the northern hemisphere, and is much more diffuse in the southern hemisphere.

12 The CO_2 cycle: seasonal polar caps

A mantle of carbon dioxide snow

As spring begins in either hemisphere, the return of daylight after months of polar night reveals a magnificent sight: from the pole down to latitude 50°, the Martian surface is covered with a brilliant icy coat which, as spring takes hold, gradually recedes polewards. These seasonal polar ice caps were first observed in the seventeenth century, and astronomers of the time had no doubts: here were

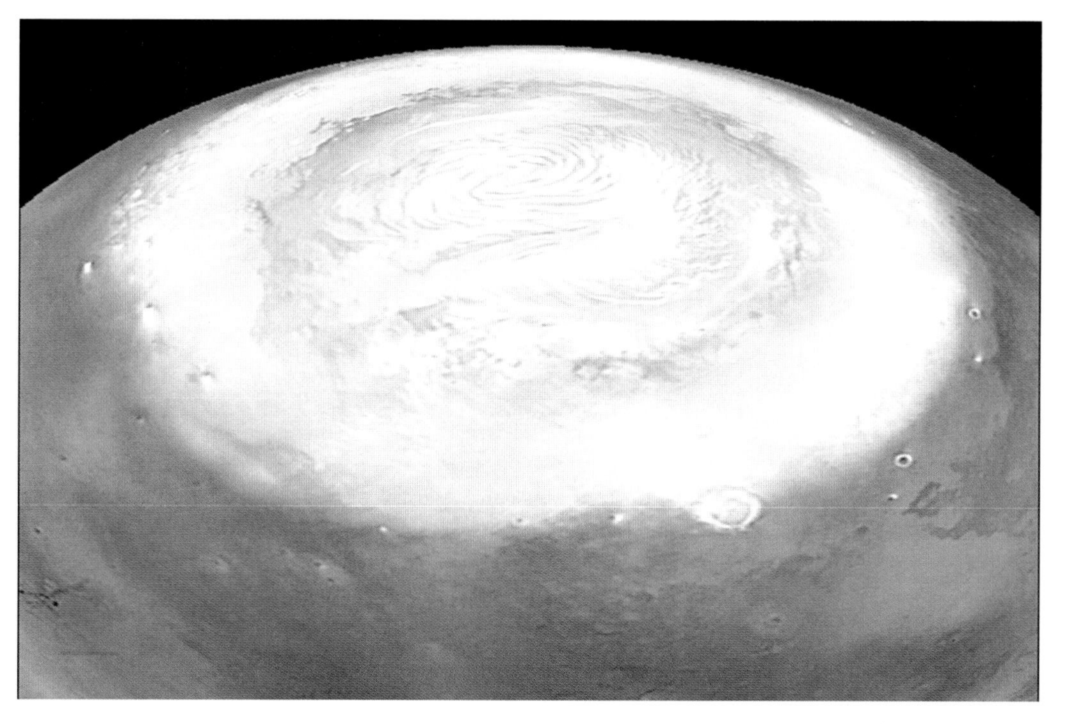

The north polar cap in mid-spring. This is a photo-mosaic of images taken at around 14.00 local time by Mars Global Surveyor in July 2002. In reality, part of the cap would always be on the night side of the planet. The seasonal caps consist of a layer of carbon dioxide ice, 10 cm to 1 m thick, which has accumulated during autumn and winter, and progressively returns to the gaseous state (sublimation) in spring. On this image, the cap is only 3000 km wide, having shrunk from its midwinter diameter of 5000 km.

The seasonal polar caps also contain a small amount of water ice in crystalline form. At the edge of the caps, in spring, this water ice takes longer than the carbon dioxide to sublimate. This explains why the OMEGA mapping spectrometer on Mars Express showed that the CO_2 cap is surrounded by a belt of water-based frost a few hundred kilometres across. (Courtesy NASA, JPL, Malin Space Science Systems.)

Beware – when is an ice cap not an ice cap?

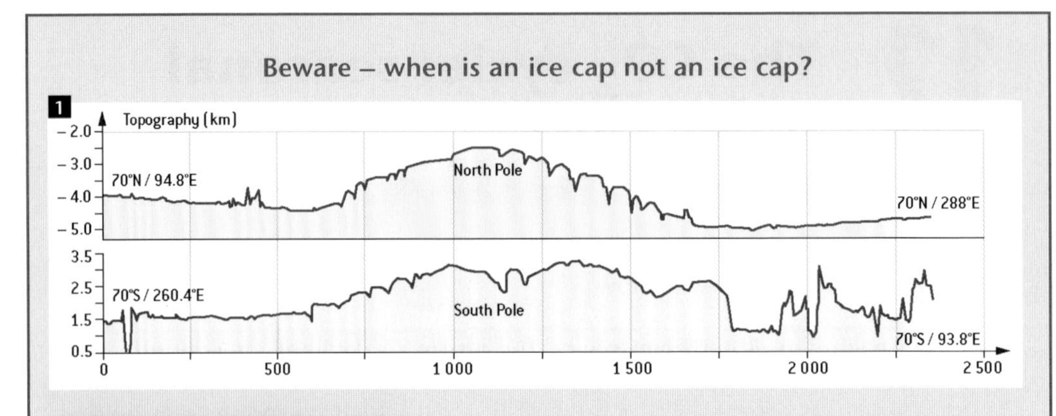

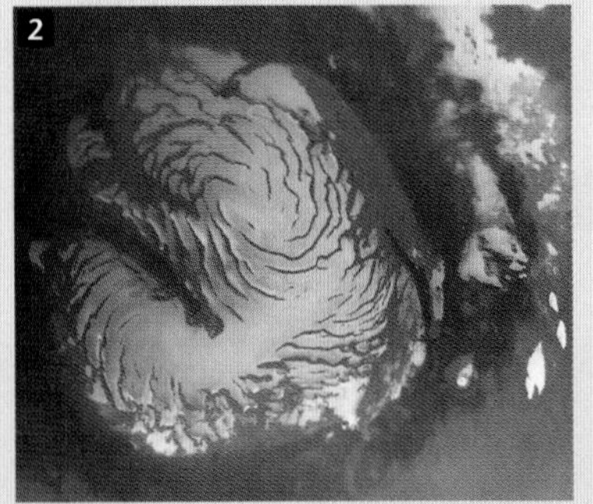

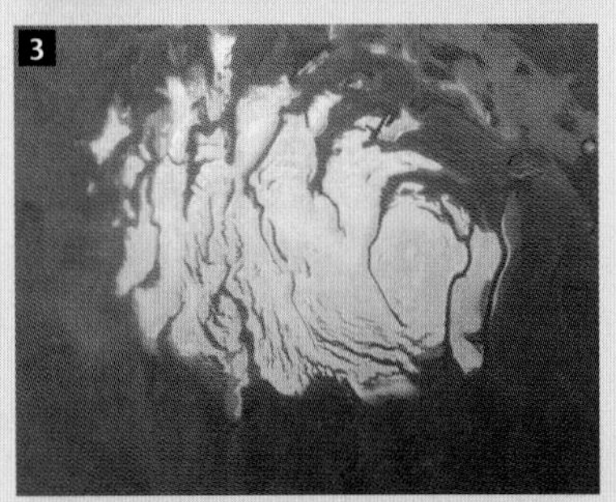

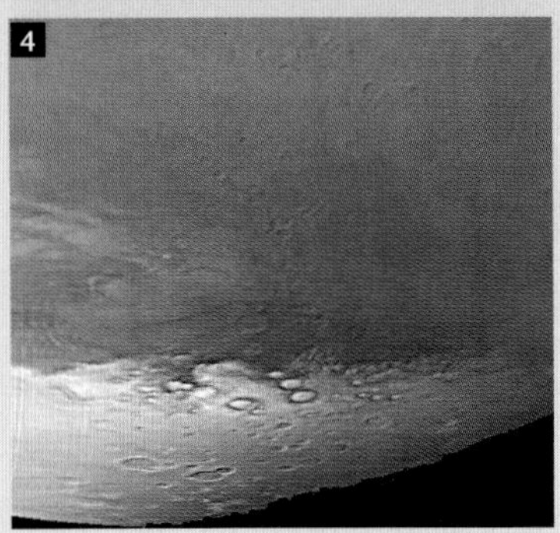

On Mars there are a number of distinct objects which may be labelled 'polar caps'.

1. At both north and south poles, there is a deposit of ice, sediments and dust which has accumulated to a depth of several thousand metres, and is about 1000 kilometres in diameter (see pages 179–181).

2. Near the north pole, the deposits have been almost entirely covered with a layer of relatively pure water ice, which interacts with the atmosphere (see pages 149–151). (Courtesy NASA, MOLA Science Team.)

3. Near the south pole, the deposits are partly covered with 'dry' sediments and with a small permanent cap of solid and very reflective carbon dioxide ice. This cap is about 400 kilometres across and about 10 metres thick, and it sits upon a substrate of water ice (see pages 164–166). (Courtesy NASA, JPL, Malin Space Science Systems.)

4. Each autumn, winter and spring, all the polar regions and their overlying deposits are covered with a seasonal layer of carbon dioxide ice a few tens of centimetres thick (as described in this section). (Courtesy NASA.)

deposits of snow and frost just like those found on Earth. Only in 1966 did their true nature become apparent, when analysis of data from the first Mars probe, Mariner 4, revealed that they were made of frozen carbon dioxide, dry ice. What is happening here is that, in the polar regions, during autumn and winter, the temperature falls until the point of condensation (solidification) of carbon dioxide is reached. On Mars, this occurs at around –125°C; the value depends on the atmospheric pressure. So it is the atmosphere itself which is condensing, covering the ground with a sheet of snow and carbon dioxide ice. In spring, the solid carbon dioxide will again sublimate (i.e. pass directly, without liquefaction, to the gaseous state).

In the northern hemisphere, the seasonal CO_2 layer will progressively shrink as spring wears on, and by the beginning of summer, it will have completely gone. As it disappears at the north pole, the permanent water-ice cap is revealed, to remain visible until the carbon dioxide returns the following autumn.

In the southern hemisphere, the winter retreat is less regular, and areas of CO_2 will disappear in some localities, while others persist at the same latitude. Near the south pole is a region where the CO_2 never completely sublimates. This 'residual' cap represents a permanent reservoir of solidified atmospheric gas, a phenomenon without parallel on Earth (see pages 164–166).

13 The CO_2 cycle: an atmosphere which solidifies

CO_2 snowfalls in the polar night

It is April 1998. Northern winter on Mars. After long months of manoeuvring, Mars Global Surveyor finally aims all its instruments at the planet's surface. The MOLA laser altimeter measures the topography to within a few metres. To everyone's surprise, it discovers structures more than 10 kilometres high, scattered below in the polar night, which nobody has ever detected before! It is soon realised that these are in fact clouds, which, it seems, exist only in the darkness. These clouds, and there is nothing like them on Earth, are very probably made of solid carbon dioxide, as has been predicted in theoretical models of the atmosphere. At first sight, it does not seem difficult to describe how they come about: as the temperature falls to the point of condensation of carbon dioxide (in Martian low-pressure conditions, –125°C), the atmosphere solidifies and particles form, aggregating into clouds.

However, the finer details of this process are not so easy to discern. These clouds are very different from their normal Martian or Earthly water-based counterparts. In water clouds, the size of the droplets is limited by the quantity of water vapour available in the atmosphere. In the case of carbon dioxide clouds, there are unlimited amounts of CO_2 to draw upon, since it is the major component of the Martian atmosphere. So when a section of that atmosphere is

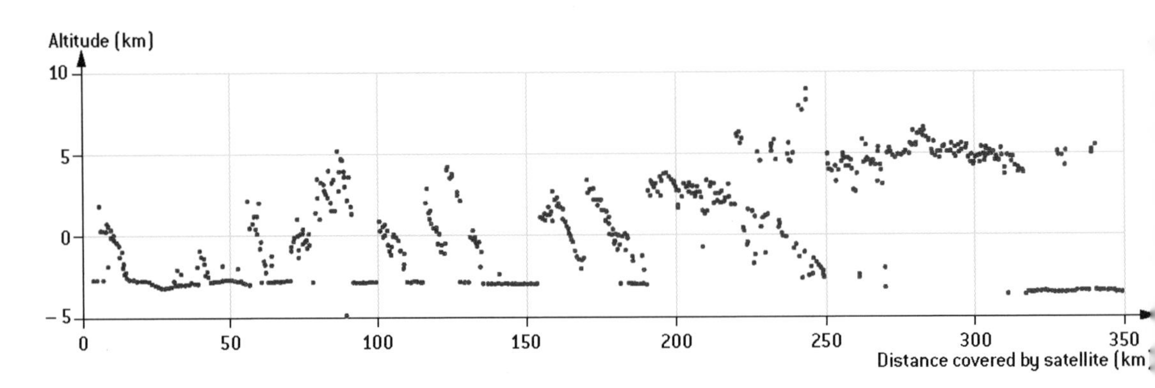

Detecting carbon dioxide ice clouds during the northern polar night, with the MOLA laser altimeter.

rapidly cooled, for example when the wind encounters relief and moves upwards into colder layers, the Martian surface experiences veritable CO_2 blizzards. The falling 'snowflakes' complement the large amounts of carbon dioxide ice which has condensed out directly onto the surface.

Generally, about 30% of the atmosphere is trapped in either the northern or southern seasonal polar cap. This condensation-sublimation phenomenon has global repercussions. First, it engenders a CO_2 flux, moving out of the sublimating cap towards the condensing cap, contributing significantly to Martian atmospheric circulation. Second, it causes seasonal variations in the total mass of the atmosphere, affecting the planet's surface pressure (see figure below).

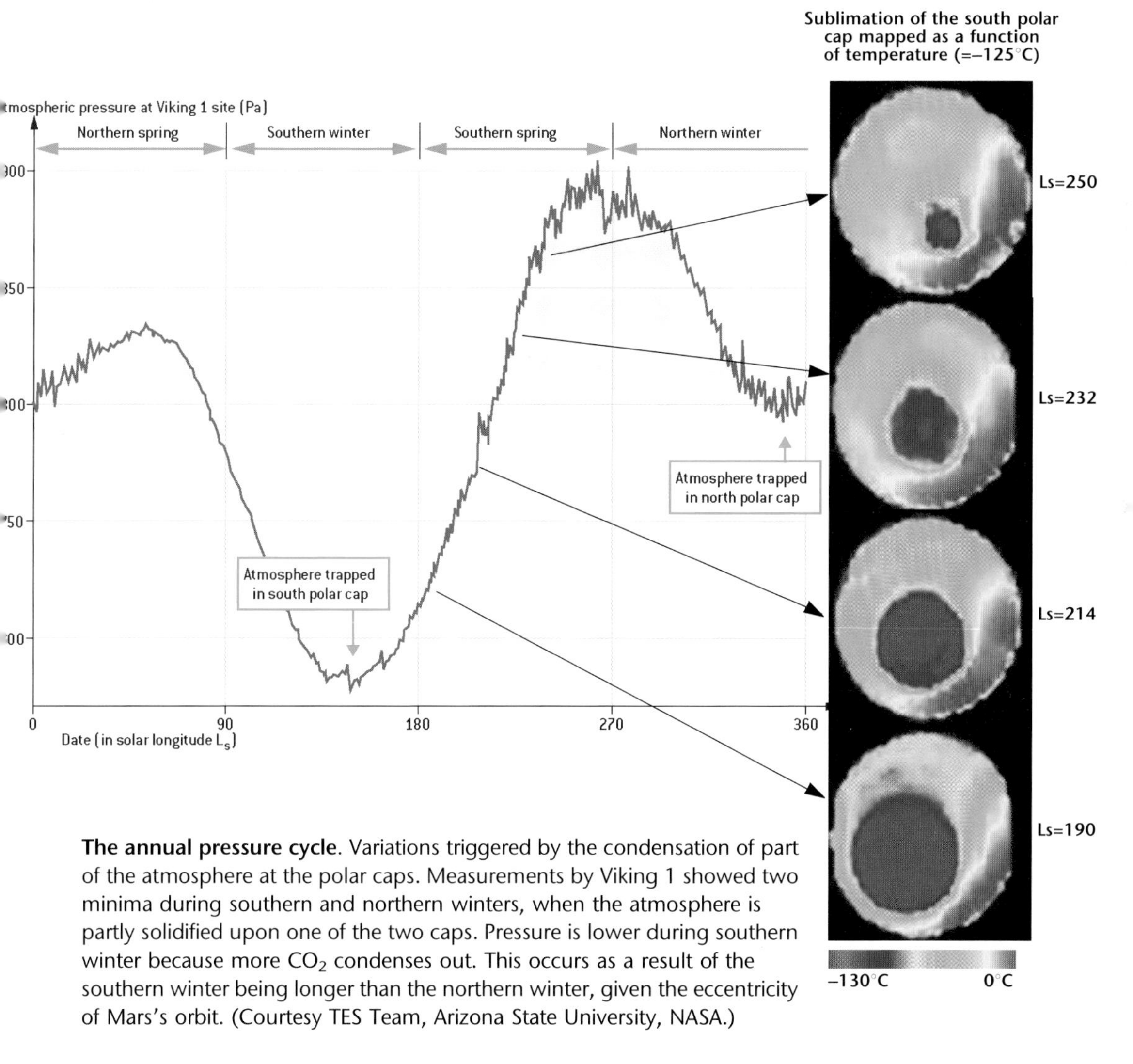

The annual pressure cycle. Variations triggered by the condensation of part of the atmosphere at the polar caps. Measurements by Viking 1 showed two minima during southern and northern winters, when the atmosphere is partly solidified upon one of the two caps. Pressure is lower during southern winter because more CO_2 condenses out. This occurs as a result of the southern winter being longer than the northern winter, given the eccentricity of Mars's orbit. (Courtesy TES Team, Arizona State University, NASA.)

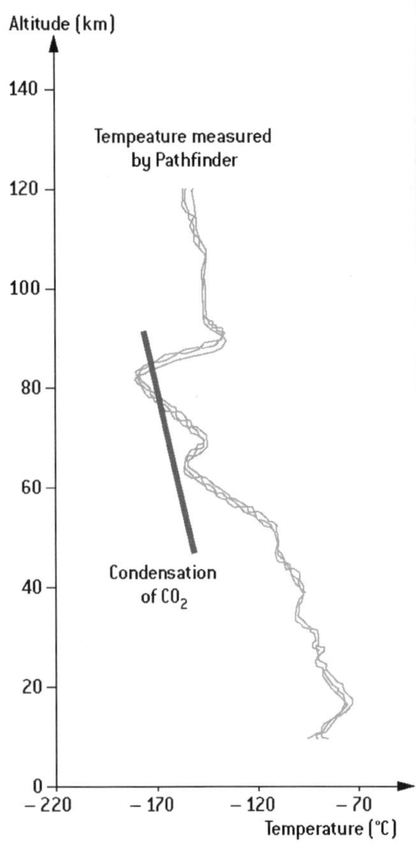

CO_2 ice clouds everywhere on Mars? It is probable that carbon dioxide clouds form not only during the polar night. This white layer (above), detected at an altitude of 100 kilometres by Mars Global Surveyor, could be composed of fine particles of solid CO_2 in suspension. (Courtesy NASA, JPL, Malin Space Science Systems.) Temperature profiles (left) recorded by the Pathfinder probe in 1997 and measured by the Spicam instrument aboard Mars Express, show that the high-altitude atmosphere may reach the condensation point of CO_2 even in the tropical zones in summer. SPICAM's observations indicated that these thin clouds tend to form in places where temperatures are very low. The true nature of these clouds is difficult to determine because they are often too thin to be directly studied by spectrometers. The OMEGA imaging spectrometer on Mars Express has detected perturbations in the radiation emitted by the atmosphere, which seem to be attributable to the presence of carbon dioxide ice particles in suspension.

14 The CO_2 cycle: the vagaries of carbon dioxide ice

Solid CO_2: a natural – yet extraterrestrial – material

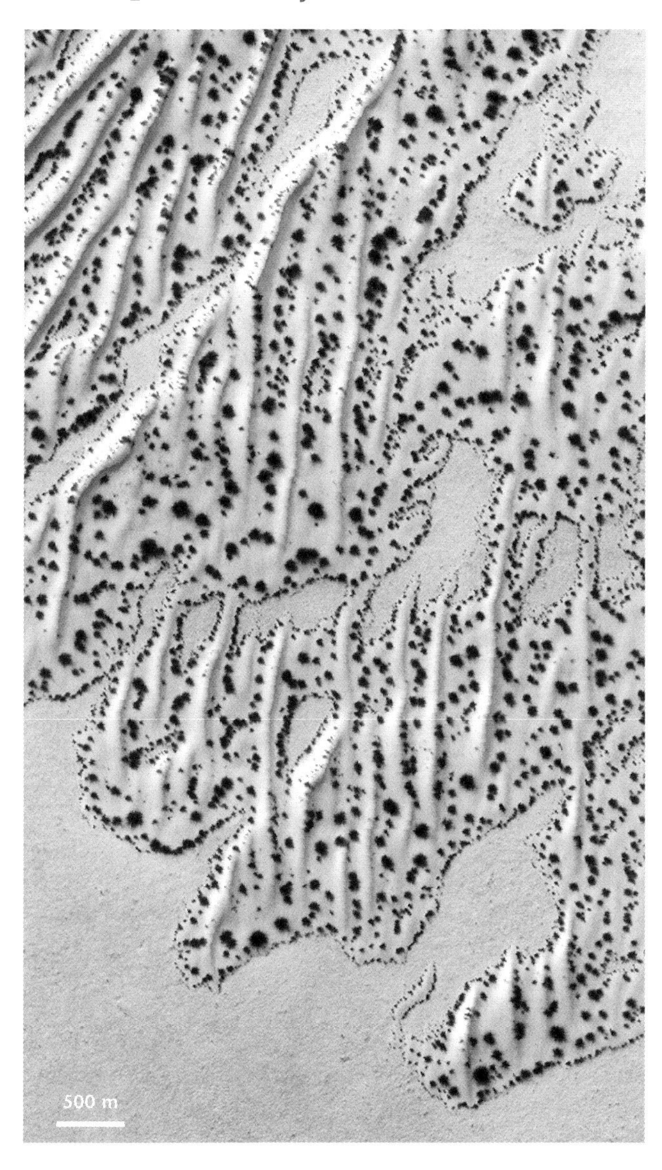

Carbon dioxide snow sublimates in the spring on the dunes of the southern hemisphere. The dark spots on the ice are thought to be composed of sand and dust carried up in powerful jets of carbon dioxide gas that form as the ice starts to sublimate in early spring, when the Sun rises again above the horizon (see next page). (Courtesy NASA, JPL, Malin Space Science Systems.)

Every spring brings violent eruptions to the south polar ice cap of Mars, according to researchers interpreting observations by NASA's Mars Odyssey orbiter. In this impression, dust-laden jets shoot into the polar sky. (Artist's impression courtesy Arizona State University, Ron Miller.)

Beneath the spring Sun, the snowbound landscapes of Martian high latitudes are certainly among the most beautiful and exotic to be found anywhere in the solar system. Satellite images of the polar caps have revealed strange structures whose formation and evolution remain mysterious. It is difficult to picture what the mantle of ice would look like seen from close to. Martian snow and ice, unlike their Earth equivalents, probably contain very few trapped 'air' (CO_2) bubbles on condensation, since it is the 'air' itself which is condensing. It is also likely that carbon dioxide ice forms very dense, homogeneous and almost transparent layers in some places. However, as the spring Sun shines, any little speck of dust locked into the transparent ice will absorb radiation, warm up and possibly surround itself with a bubble of gas as the ice sublimes around it. Beneath the ice, the rocky ground may also be warmed by the Sun, subliming the ice from below. In some areas, ice in springtime may well be in a state of 'levitation', riding on a cushion of 'air' under pressure. If a crack appears in the layer of ice, forces of aspiration under the ice may cause a veritable geyser of gas and dust to surge out onto the surface of the ice cap.

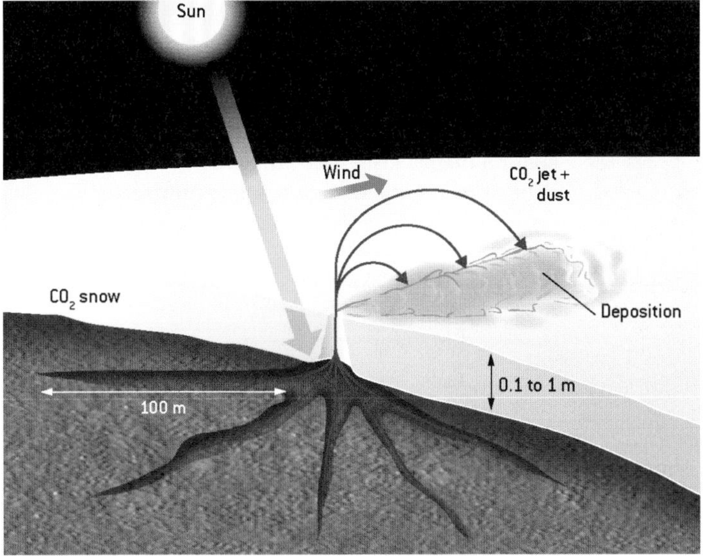

Polar 'spiders'. In certain areas near the south pole, in spring, 'geysers' of carbon dioxide and dust appear to burst out of the cap. They can be identified by the dust which they leave downwind in their vicinity (a). This gas, under pressure, probably originates in the sublimation of ice below, and is forced out by a warming of the rocky ground absorbing solar radiation through the transparent ice. As the gas surges forth, its motion beneath the ice seems to be sufficiently violent to erode the surface, and draw with it mineral particles, for distances of tens of metres. Since local relief probably determines the locations of these geysers, this kind of erosion happens each year at the same spots. As the years succeed each other, erosion patterns in the form of stars are created in the rocks and the soil in these areas. These figures may be discerned beneath the ice around the locations of geysers active in the spring (b). They appear more clearly in summer, when the carbon dioxide ice has completely disappeared. (Figure after S. Piqueux). (Courtesy NASA, JPL, Malin Space Science Systems.)

15 The CO_2 cycle: the residual southern polar cap

A reservoir of frozen atmosphere

Although all traces of polar CO_2 ice disappear in the northern hemisphere as summer comes, near the south pole, the seasonal sublimation of CO_2 snow uncovers a residual deposit of solid carbon dioxide, in equilibrium with the atmosphere.

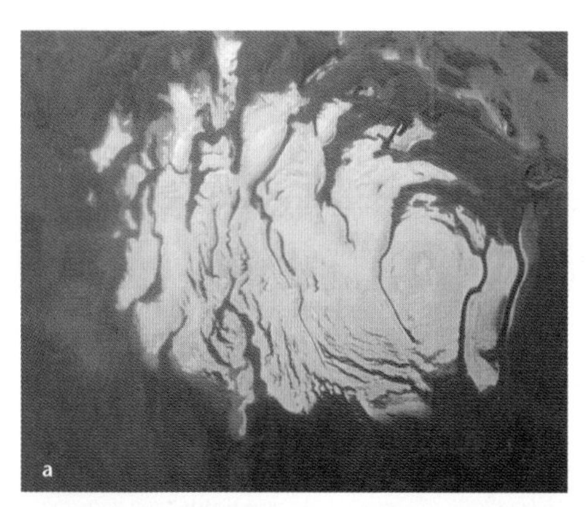

The residual CO_2 cap near the south pole (a) is 400 km across, and extremely reflective and cold (–130°C). (Courtesy NASA.) High-resolution images from Mars Global Surveyor have revealed relief unlike any other in the solar system. The cap is the result of the physical phenomena involved in the sublimation of solid CO_2 in an atmosphere of almost pure CO_2. The three photos below (b, c and d) show layers about 5 metres thick. Once sublimation begins at one point in the layer, it tends to propagate, either radially or in a preferred direction, forming patterns of circles or parallel grooves. (Courtesy NASA, JPL, Malin Space Science Systems.)

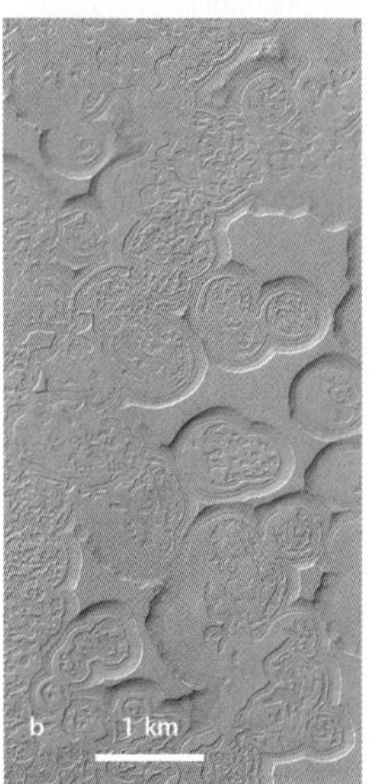

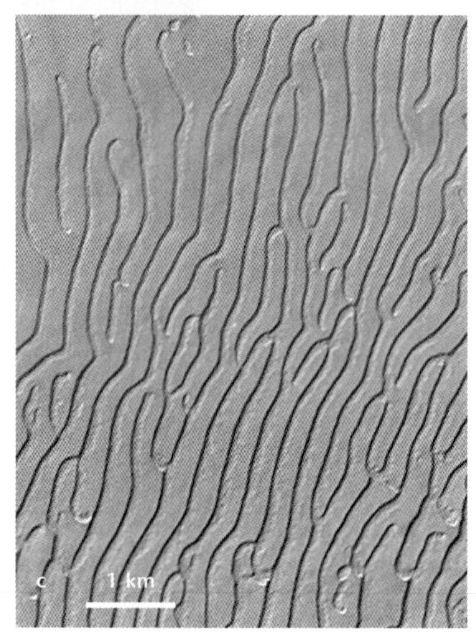

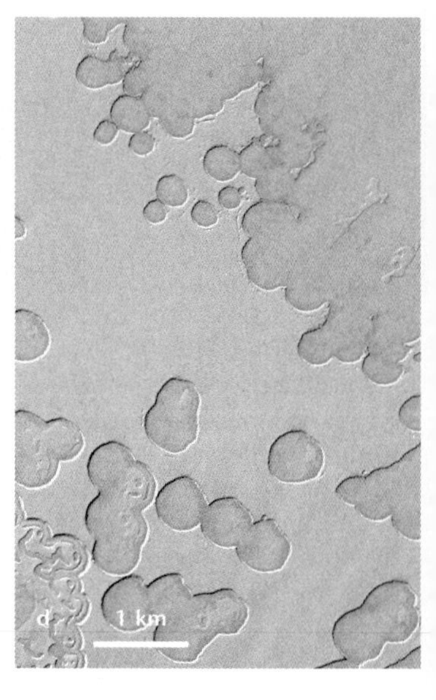

Enigmatically thin

This perennial deposit puzzles planetologists, because it seems to be astonishingly thin: just a few metres at most, or the equivalent of a few years' seasonal condensation. In the long term, such a 'glacier' should either disappear or, having accumulated much more ice, grow into something much more substantial. It does not seem possible that the natural evolution of the ice cap should somehow fortuitously hover between these two alternatives, unless some unknown feedback is acting to stabilise its fragile equilibrium. Another explanation is based on the idea of the residual southern cap forming periodically as climatic variations in the CO_2 cycle dictate, these variations taking decades to occur. The existence of such phenomena is however still a matter of speculation.

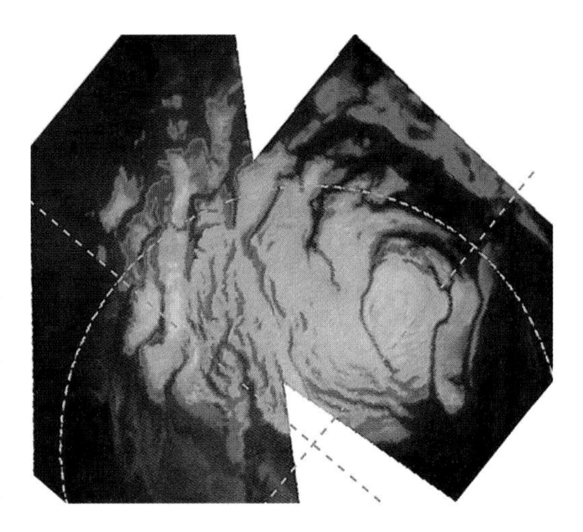

A thin layer of carbon dioxide ice on a water-ice substrate. This (false-colour) montage is made up of images obtained during southern summer by the OMEGA mapping spectrometer aboard Mars Express in 2004. OMEGA can distinguish carbon dioxide ice from water ice by infrared spectroscopy, and revealed that the 'perennial' (pale pink) carbon dioxide ice layer, lying on the (blue) water-ice, seems to be no more than a few metres thick. (Courtesy ESA, OMEGA.)

Why is a residual ice cap present only in the southern hemisphere?

The fact that there is no permanent CO_2 cap at both poles is not surprising, because such a cap will form where CO_2 ice is at its most stable, e.g. in the coldest place. Now Mars' current orbital eccentricity means that its southern summer, though warmer, is shorter than the summer in the north. This could lead to ice being more stable at the south pole. This is a situation which evolves over a 25,000-year period: we can imagine the opposite scenario, tens of thousands of years ago, with a northern ice cap covered in CO_2 while its southern equivalent was a source of water vapour.

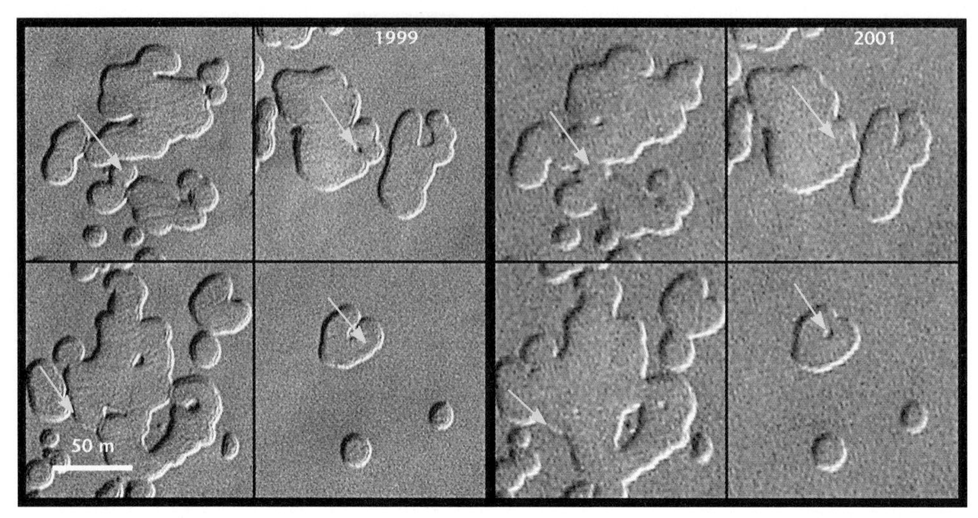

An ice sheet in the process of sublimation? A comparison of images from Mars Global Surveyor taken in 1999 and 2001 reveals the process of sublimation in part of the upper layer of the residual southern CO_2 cap. In the photos, the retreat of edges and the disappearance of internal features (arrowed) are clearly visible. The continuation and generalisation of this trend across the whole cap would suggest that the mean pressure of the Martian atmosphere is gradually increasing. (Courtesy NASA, JPL, Malin Space Science Systems.)

16 Climate change: when Mars rocks on its axis

The orientation of the planet relative to the Sun determines its climate

The climate of a planet is regulated by variations in its orbital parameters, and in particular by its obliquity (the angle between the axis of rotation and the perpendicular to its orbital plane). In the case of the Earth, the oscillations in its obliquity, though minimal (since the Earth is stabilised by the Moon), have played an important part in the onset of glaciations. Recent calculations have demonstrated that Mars has undergone a chaotic series of variations in its obliquity, with a 'pseudo-period' of about 100,000 years and extreme values of as little as 0°, and of more than 60°. Needless to say, Mars (current obliquity 25.2°) has a history of considerable climatic variations. In what way?

When the obliquity is small (0° to 20°)...

Seasons are less marked. The poles, where the seasonal CO_2 ice caps are less extensive, receive on the average less energy, since the Sun is always low in the sky. Polar temperatures are lower. As a consequence, it is probable that the fraction of the atmosphere constantly held as CO_2 ice (which is at present at the south pole – see previous pages) increases. The atmosphere is therefore less dense

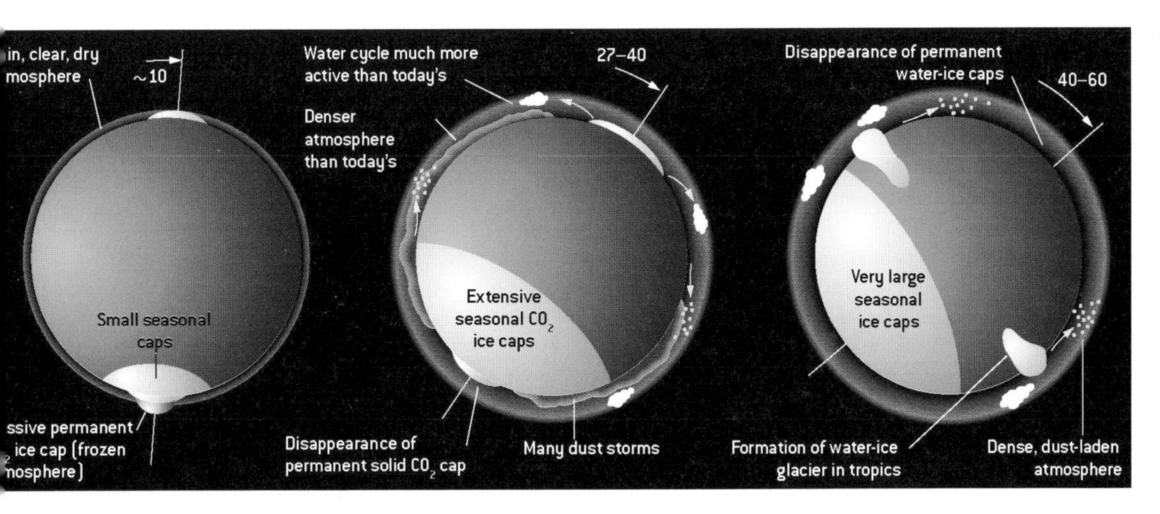

The effects of obliquity on the Martian climate. (a) Low obliquity (0° to 20°); (b) high obliquity (27° to 40°); (c) very high obliquity (40° to 60°).

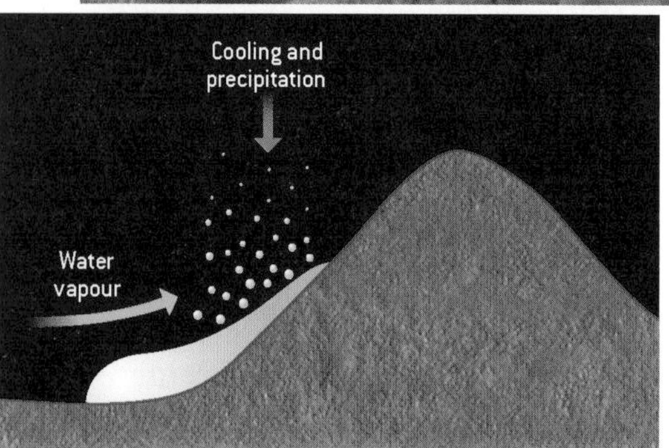

Glaciers in the tropical zones of Mars, a few million years ago. Formations similar to certain deposits left by large glaciers on Earth are found on the western flanks of the great Martian volcanoes in the Tharsis region, and adjacent plains (shown in yellow on the map). These vast Martian glaciers have by now almost completely vanished.

Nevertheless, Mars Express' HRSC camera discovered, near the scarp of Olympus Mons, a so-called 'rock glacier' (a). (Courtesy ESA, DLR, FU Berlin (G. Neukum.) Composed of ice covered with rocks and sediments, this glacier resembles those in the 'dry valleys' of the Antarctic (b), areas which, like Mars, are particularly cold and dry. (Courtesy D. Marchant.) How could such glaciers have formed on a planet where surface ice is unstable except at the poles? Computer simulations of the Martian climate at times of high obliquity show that the atmosphere was then much richer in water vapour than it is today. Impelled by westerly winds onto the flanks of the volcanoes, this moist atmosphere was abruptly cooled. Ice condensed out and fell to the surface. It is estimated that the rate of accumulation was something like a few centimetres per year, enough to form a glacier over thousands of years. Now, most of the water has evaporated; but, at the heart of the rocky glacier, a small amount of ice may remain beneath its protective layer of rocks and dust.

than is observed today. Models suggest that, in these conditions. water and dust cycles would be much less active, and the thin atmosphere would remain clear.

When the obliquity is large (more than 27°)...

Seasons are more obvious. Greater annual mean insolation at the poles means that there will be more warmth at depth. Trapped CO_2 ice or carbon dioxide 'adsorbed' in the pores of the Martian sub-soil are released, until the reservoir is exhausted. The capacity of this reservoir is unknown, but we can envisage an atmosphere up to twice as dense as the present one. Circulation within this atmosphere would intensify as seasons became more marked. Great quantities of dust could be raised, and maintained in suspension. In summer, the polar water-ice reservoir would be considerably warmer, and a great volume of water would be released; the hydrological cycle involving clouds and frost would be much more dynamic than that which we observe today.

The formation of glaciers when obliquity is very large (more than 40°)

Computer simulations of the Martian climate suggest that warming of any ice left at the poles would liberate such a large amount of water into the atmosphere that, in certain areas, water vapour would condense and precipitate out much more readily than it could sublime. In these areas, ice would therefore have accumulated (as long as there was still a source at the pole) and might even have formed glaciers. The remains of these glaciers have indeed been found where the models indicate they once existed, on the flanks of the great volcanoes of Tharsis and to the east of the Hellas Basin (see above, and next page).

17 Climate change: ancient glaciers on Mars

Structures comparable to glaciated valleys on Earth

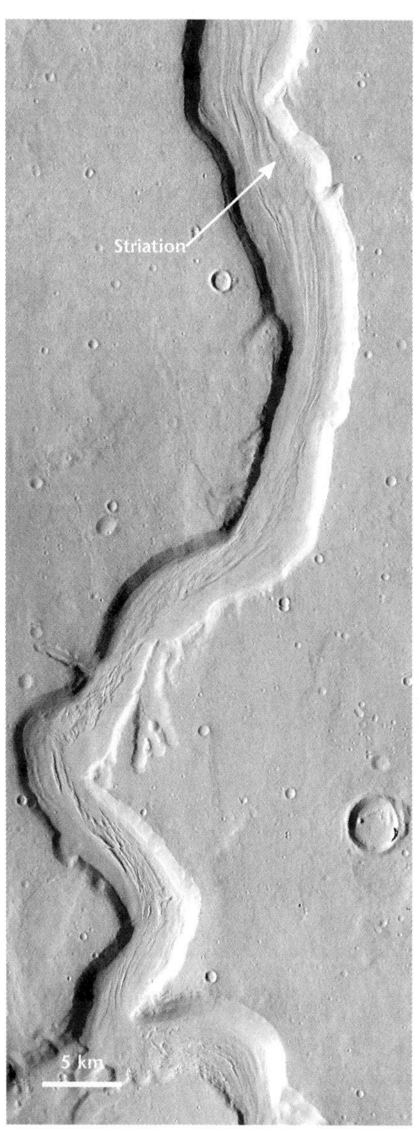

Well before computer simulations of the Martian climate predicted the formation of glaciers, as a result of fluctuations in the planet's orbital parameters and obliquity (see previous pages), geologists had identified signs of glaciers in certain specific areas. For example, on the boundaries of the cratered highlands and the lower-lying northern plains, there are long valleys 5 to 10 kilometres wide and 1000 metres deep, with a peculiar morphology. Their flat floors are scored in the direction in which the valleys run. These score marks, with their alternating ridges and crevasses, start at the valley walls and curve inwards to follow the longitudinal slope of the valleys. They seem to

A valley in Arabia Terra covered with moraine-like scour marks. This morphology is reminiscent of glaciated valleys on Earth. The area in the image measures 57 km by 23 km, with illumination from the left. (Courtesy NASA, JPL, Arizona State University.)

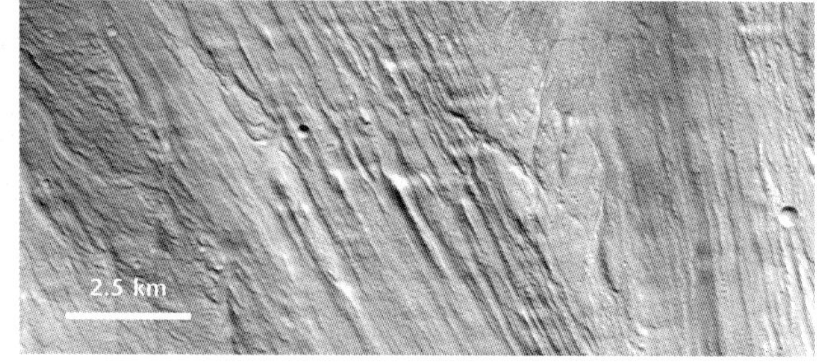

Scour marks on the floor of the outflow channel Kasei Vallis (see pages 111–113), which may have been covered in ice in the past (Mars Odyssey Themis image, 2002). (Courtesy USGS, A.L. Washburn.)

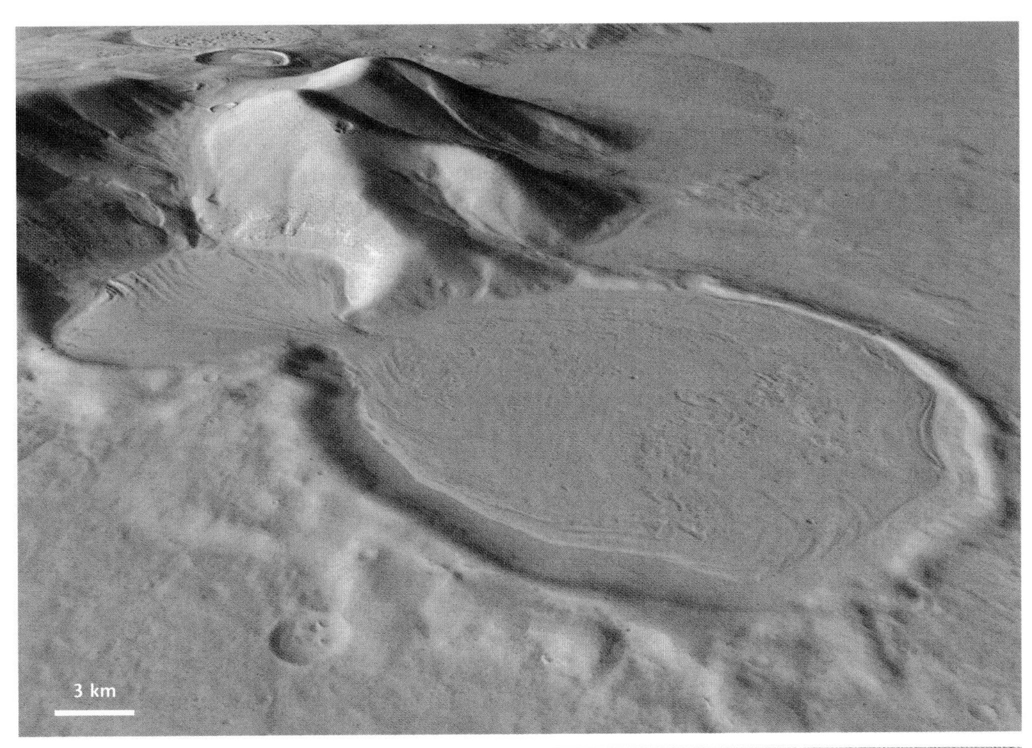

Rock glaciers to the east of the Hellas basin. The upper image is a 3-D stereo view of two craters filled with sediments, apparently very rich in ice. The ice has flowed from the smaller crater (9 km across) into the lower, larger one (16 km). (Courtesy ESA, DLR, FU Berlin (G. Neukum.) The lower image shows a 5-km 'tongue' which resembles a terrestrial rock glacier. These, and hundreds of other similar features, can be found in a comparatively small region 1000 km wide, just east of the Hellas basin. (Courtesy NASA, JPL, Malin Space Science Systems.) Climate models have indicated that the particular nature of this region was determined by relatively abundant precipitation. This would have occurred when, during a period of high obliquity, Mars' south pole was a source of water vapour in the southern summers. The topography of the Hellas basin would have perturbed the atmospheric flow, causing most of the water vapour to pass to the east of it. There, the water-laden air encountered colder air, and some of the water condensed.

be evidence of the presence of a mixture of ice and rocks which once 'flowed', in the manner of glaciers on Earth. Rock debris from the valley walls formed moraines, which were drawn along by the glaciers. In other valleys, the ice seems to have disappeared, but there are traces of scoring similar to those seen in ancient glaciated valleys on Earth. This suggests that the surface has been eroded by the passage of a layer of ice about one kilometre thick. Given Mars' lower gravity and the very gentle slope of the valleys, the ice must have been at least of that thickness to have been able to set the glacier in motion.

How could these glaciers have formed in the atmospheric conditions found on Mars? Not all of these ancient glaciers were necessarily created by precipitation from the atmosphere. We can envisage that, in some cases, water issued from an underground reservoir under pressure, and once on the surface, it flowed away and froze.

18 Climate change: surface ice at high latitudes

Above latitude 60°, a sheet of ice laid down millions of years ago extends beneath the surface

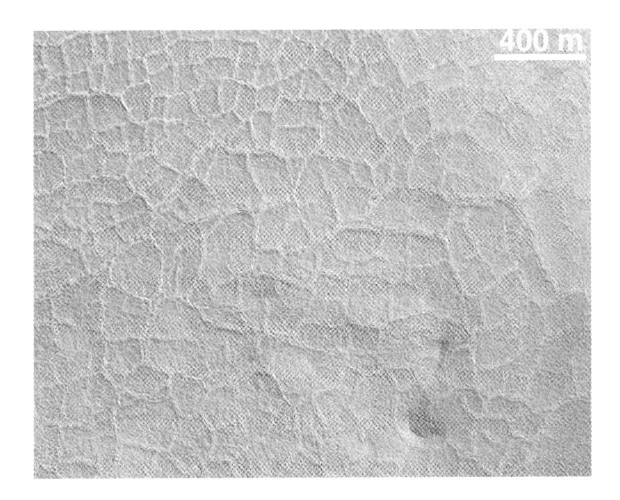

Polygons in the ice. Between latitudes 60° and 80° on the northern plains of Mars, the presence of polygonal structures on the surface, similar to those found on Earth in the Arctic, show that the soil is laden with ice, and has been patterned by the action of thermal contractions. (Courtesy NASA, JPL, Malin Space Science Systems.)

High-resolution images from Mars Global Surveyor, and data from the gamma-ray spectrometer on Mars Odyssey, have recently revealed that, less than a metre below the Martian surface at latitudes above 60°, there exists a considerable amount of frozen water.

In October 2001, Mars Odyssey, which had recently arrived to accompany Mars Global Surveyor in orbit around the planet, carried new types of detectors (see Part 5). One of these was the gamma-ray spectrometer, working in tandem with neutron detectors (neutrons are particles which, with protons, make up the nuclei of atoms). The neutron detector was able to reveal the presence of water below the surface, by measuring the flux of neutrons from Mars. Neutrons are emitted by sub-surface atoms during their continual bombardment by energetic particles (cosmic rays) emitted by various objects in the universe. So the red planet constantly emits neutrons which may be intercepted and analysed by detectors aboard space probes. The number of neutrons radiated by a particular region can show the presence of ice just beneath its surface, because, before being released into space, these particles have to traverse the various sub-surface layers of Mars. During their journey, the hydrogen in water molecules is particularly effective at slowing them (a property of water utilised in nuclear reactors). The more ice there is near the surface, the weaker the neutron flux will be. There was general surprise when Mars Odyssey's detectors transmitted evidence that, in certain

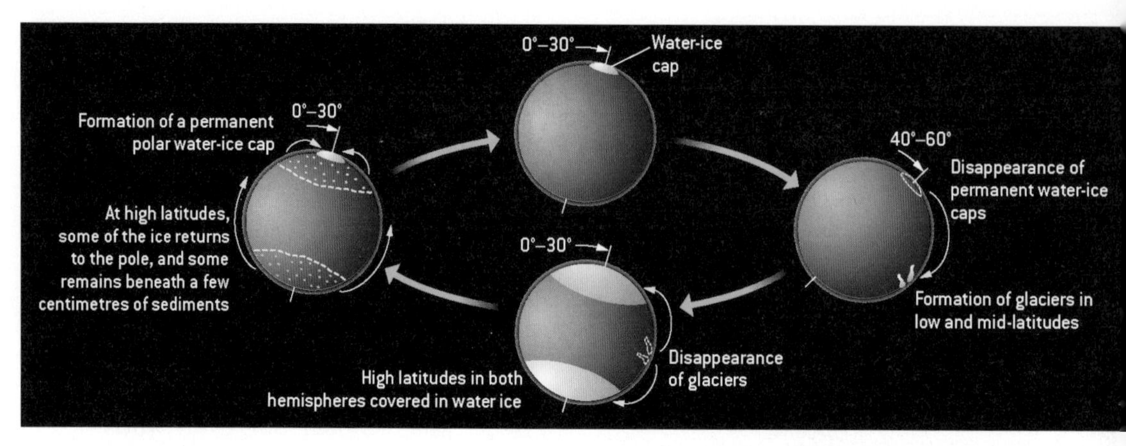

Where does Mars' icy layer come from? Some researchers support the hypothesis that the ice appeared by simple diffusion into the pores of the upper sub-surface material. There are many, however, who think it more realistic, given the concentrations of ice detected, that an icy layer must once have covered Mars at mid- and high latitudes. The scenario described here is based on climate simulations which show that, when the obliquity of Mars approaches low values such as we see today, after a period of high obliquity, water which has accumulated in tropical glaciers does not return directly to the pole. It spreads across each hemisphere in the zone above 60°, forming a layer of ice and dust a few metres thick. Once the source in the tropics dries up, the upper part of this layer will sublime, and the water will make its way to the pole. It would leave behind a blanket of dust thick enough to limit the loss of the rest of the ice.

regions, there are large quantities of water frozen into the surface to a depth of one metre. Results indicate that there are concentrations of ice representing more than 70% of the sub-surface material at latitudes above 60°. Here, there must be an actual layer of ice just a few centimetres below the dry surface.

Below latitude 40°, there may be ice buried more than a metre down, but at such a depth, Mars Odyssey is unable to detect it.

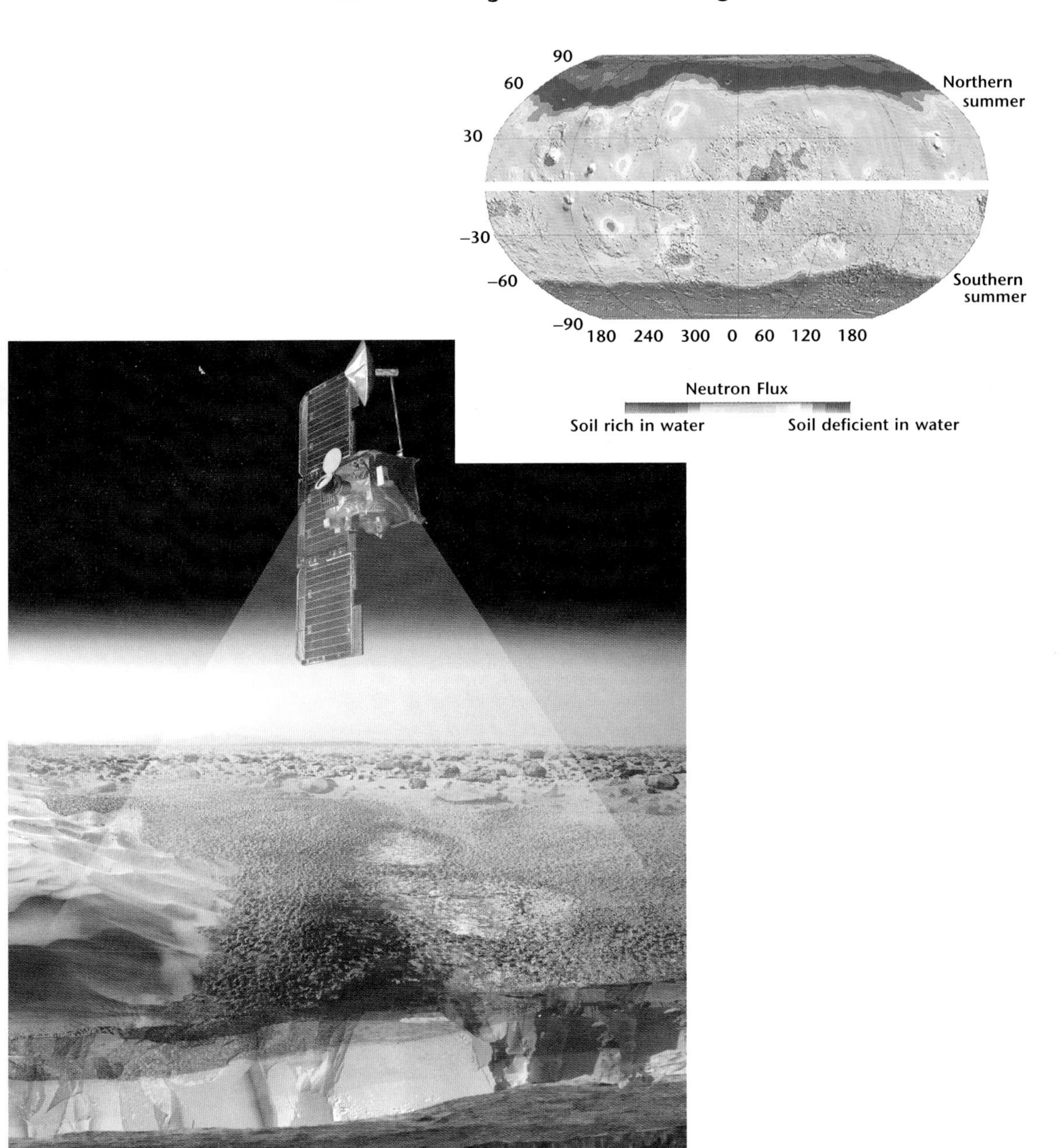

Map of neutron flux levels on Mars. These measurements, by the neutron detector aboard Mars Odyssey (above), show the water-ice content of the soil down to one metre. In blue and violet are zones of high water-ice content, and in green, yellow and red, zones of low content. These measurements can be carried out in the polar regions only in summer, when the seasonal carbon dioxide ice cover has disappeared (see pages 155–157). (Courtesy NASA, JPL, University of Arizona, Los Alamos National Laboratories.)

19 Climate change: intermittent water flows

Traces of liquid water just a few hundred thousand years old

Planetologists were in agreement that the scenario involving liquid water flowing across the surface of Mars had indeed happened, but in Mars' very distant past, billions of years ago. Then, in 1999, images from Mars Global Surveyor, broadcast around the world, seemed to contradict this assumption. The probe's MOC camera revealed, at high and mid-latitudes, systems of gully-like structures with an apron of debris at the end, probably created by flowing water in geologically recent times: perhaps a few million years ago, perhaps much later. This discovery reanimated the debate: has water flowed recently on Mars? Is it still flowing today, or could it be the result of climatic changes

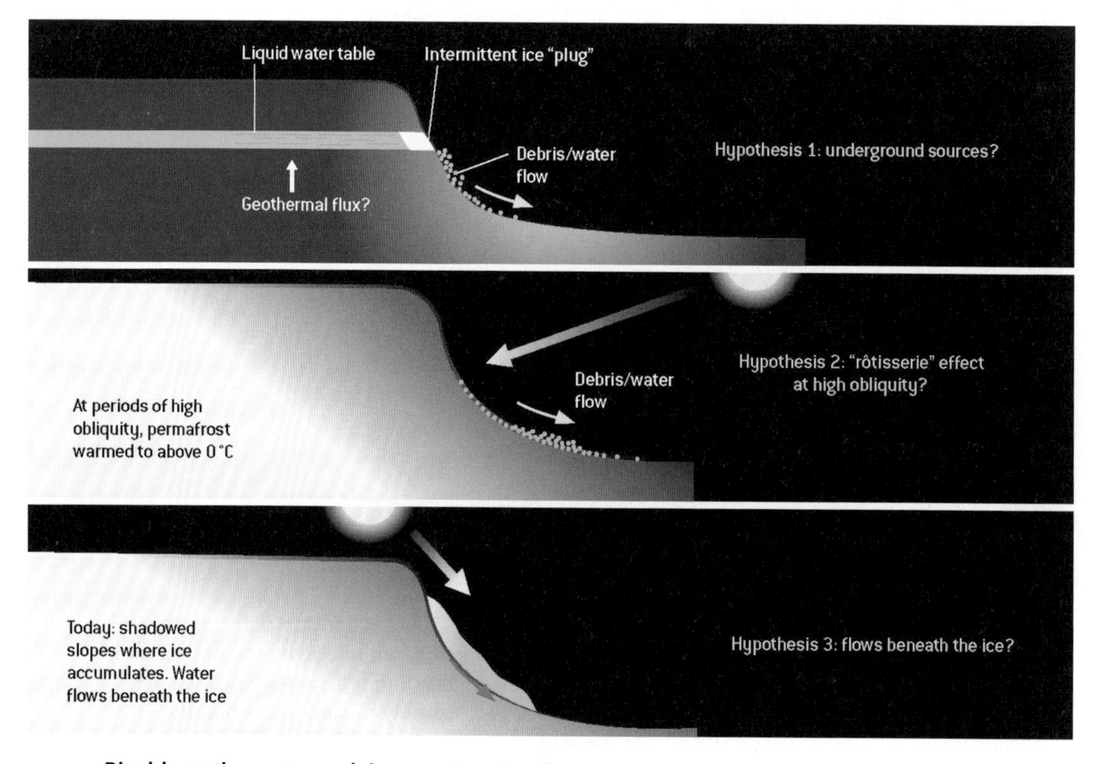

Rival hypotheses to explain recent water flows.

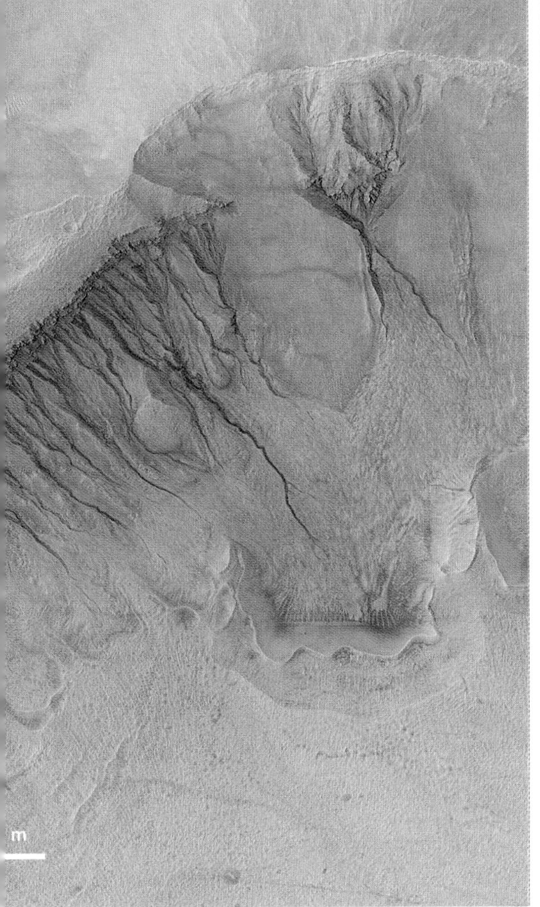

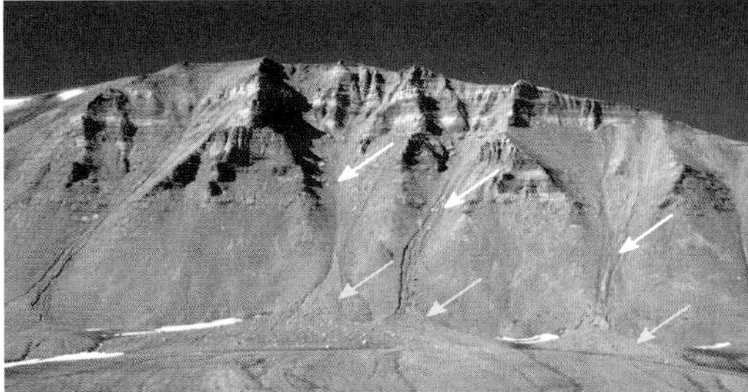

Similar gullies exist on Earth in periglacial regions, where there is permafrost. For example, these 500-metre-long debris flows in Greenland have been triggered by the thawing of ice or snow on the surface, when rocky material strewn about the slope becomes waterlogged (white arrows). After flowing down the slope, the debris spread out to form characteristic cones (yellow arrows) or low lateral ridges known as *levées*. All these features can be found on Mars. (Courtesy F. Costard.)

High-resolution images from Mars Global Surveyor show gullies cut into the internal walls of some impact craters. These features seem to originate at the lip of the crater wall. Most of the gullies occur on pole-facing walls (north-facing in the northern hemisphere, south-facing in the southern hemisphere). This example is at 39.0° S, 166.1° W. (Courtesy NASA, JPL, Malin Space Science Systems.)

orchestrated by variations in the planet's orbital parameters? Since 1999, sites with gullies have been carefully monitored by the MOC camera. New gullies have not been detected, but in two pre-existing gullies, MOC found that a light-toned material had flowed down. This enigmatic discovery attracted considerable attention once again.

What is the origin of these features caused by flowing water?

It is not easy to explain how water can have flowed in regions where the temperature is currently well below zero and the pressure will not in theory allow liquid water to exist. Three hypotheses are at present 'in the frame'.

Sub-surface sources?

Some researchers put forward the idea that there are aquifers below ground, warmed by energy from within Mars. These could flow at the surface intermittently (first diagram, page 176). However, this scenario does not satisfactorily explain the distribution of the flow features: they seem unconnected with any volcanism. Neither does it address the fact that these features are observed on isolated peaks and even on dunes (see photo at right).

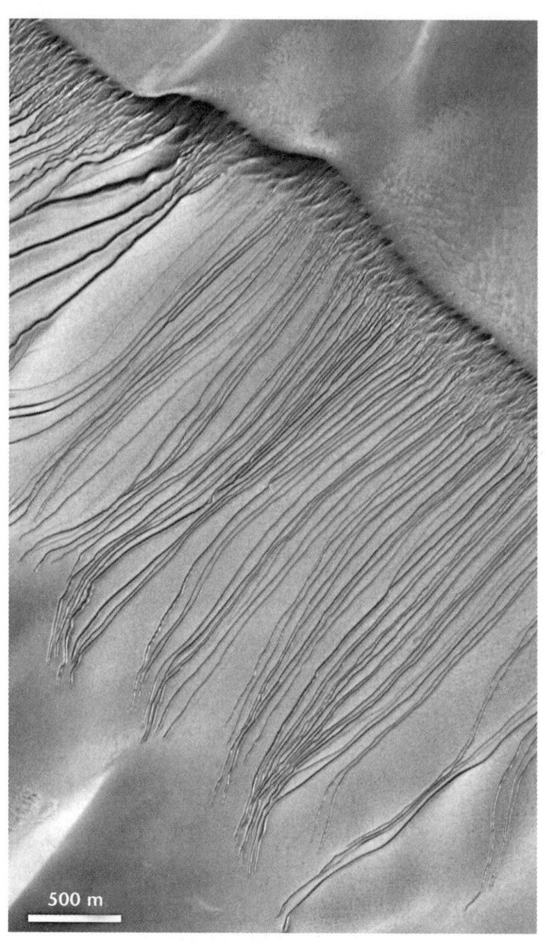

Gullies on Martian dunes. Such markings tend to contradict the theory of underground water sources (see diagrams on page 176) – there can certainly be no springs emerging from the tops of dunes! (Courtesy NASA, JPL, Malin Space Science Systems.)

A 'rotisserie' effect at times of large obliquity?

Other planetologists (among them, two of the present authors) have proposed that the flow features were created at times when the obliquity of the planet was large (see pages 167–169). Atmospheric pressure would then have been high enough for water to exist in its liquid phase, above 0°C. In summer, when the axis of rotation leaned towards the Sun, pole-facing slopes would be continuously warmed, and the surface ice or the frozen sub-surface would have melted to some depth, with consequent outflows (second diagram, page 176).

Melting below snowpacks today?

Some flow features are visible on terrains which are partly covered with deposits of ice accumulated on shadowed slopes. On this basis, some researchers have envisaged liquid water flowing beneath this protective layer of ice (third diagram, page 176). The question then is: what heat source causes this water to flow beneath the ice?

20 Climate change: polar sediments, evidence of past climates

Ice accumulated at the north pole might carry a record of the fluctuating climate of several million years

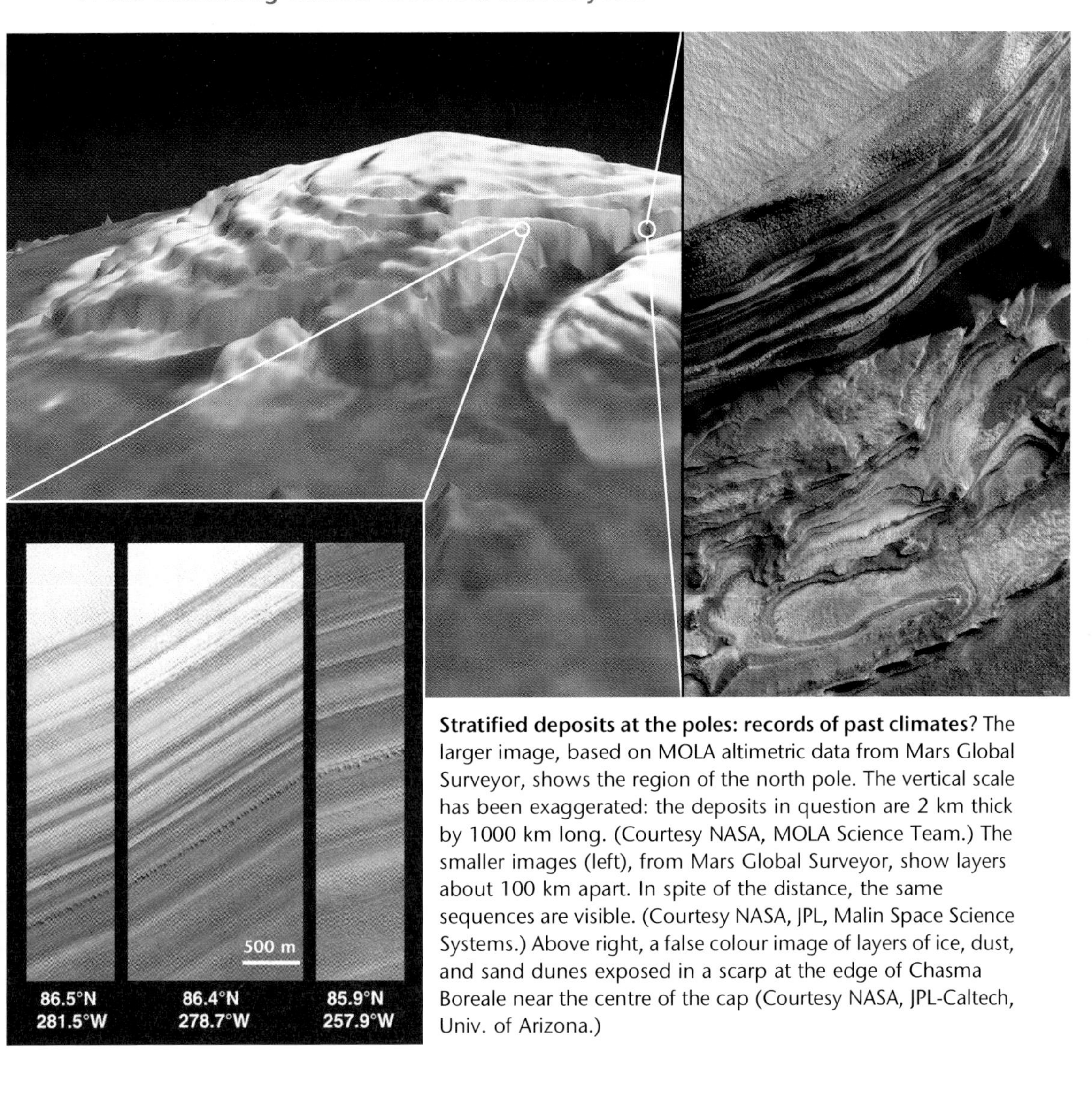

Stratified deposits at the poles: records of past climates? The larger image, based on MOLA altimetric data from Mars Global Surveyor, shows the region of the north pole. The vertical scale has been exaggerated: the deposits in question are 2 km thick by 1000 km long. (Courtesy NASA, MOLA Science Team.) The smaller images (left), from Mars Global Surveyor, show layers about 100 km apart. In spite of the distance, the same sequences are visible. (Courtesy NASA, JPL, Malin Space Science Systems.) Above right, a false colour image of layers of ice, dust, and sand dunes exposed in a scarp at the edge of Chasma Boreale near the centre of the cap (Courtesy NASA, JPL-Caltech, Univ. of Arizona.)

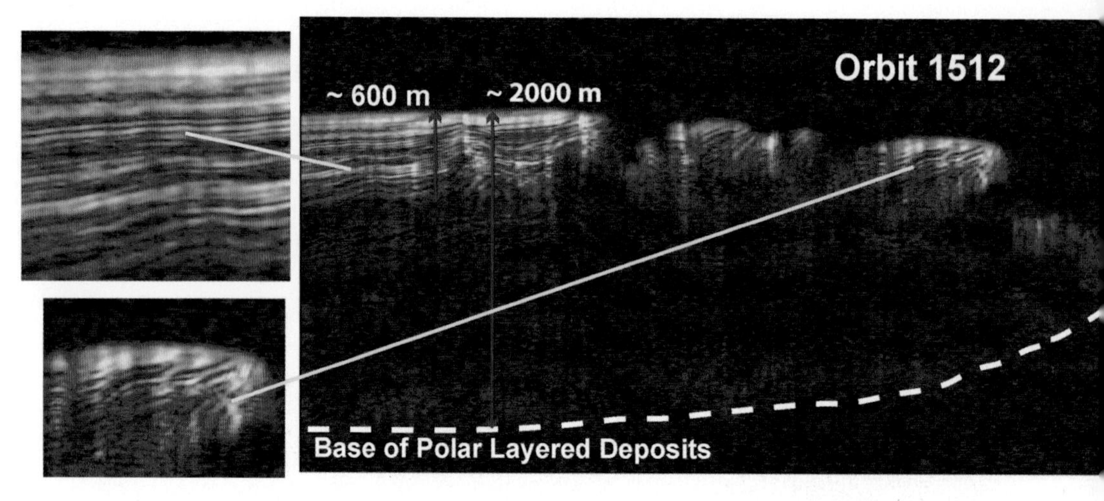

Beneath the ice. The MARSIS and SHARAD sounding radar aboard Mars Express and Mars Reconnaissance Orbiter have revealed detailed internal structure in the polar layered deposits, and discovered that the underlying rock surface is at a depth of more than 2000 metres below relatively pure ice. The image above shows a 'radargram' obtained on November 22, 2006 by SHARAD. (Courtesy NASA, JPL, ASI, ESA, Univ. of Rome, MOLA Science Team.)

The north and south poles of Mars are covered with vast deposits a thousand kilometres wide and about 3000 metres deep. At the north pole, the accumulated sediments are riven with enigmatic troughs, in a spiral pattern, upon the flanks of which can be seen the very structure of the polar cap. There are hundreds of layers less than 10 metres thick, alternately light and dark depending on the proportions of ice and dust within them. For distances of hundreds of kilometres, the same pattern of layers can be repeated. This suggests a global climatic origin, as different layers are deposited as orbital parameters vary (see pages 167–169).

So, in much the same way that ice cores from Antarctica are used to study past climates on Earth, these deposits could provide insights into the climatic history of Mars. How long a history? If we suppose that the polar cap accumulated gradually after the last period of large obliquity, supplied by glaciers further south, for example (see pages 170–175), then this cap represents five million years of frozen archives.

Ice and gypsum. The icy deposits of the north polar region are surrounded by huge, dark dune fields, whose origin remains a mystery. The OMEGA mapping spectrometer recently revealed that some of these dunes contain an unexpected mineral: gypsum (shown in red on this montage of false-colour images; the ice is white). Its concentration tails off from a point at the edge of the ice cap. Gypsum is a sulphate of calcium, and is usually formed in the presence of liquid water. How could liquid water have been present near the pole, on fairly recent terrain? The gypsum probably originates in a geological layer once modified by liquid water, at a time when the edge of the cap partially melted. This could have occurred because the ice was much thicker than it is now. The formation of pockets of liquid water beneath glaciers and polar ice sheets is also observed on Earth. (Courtesy ESA, OMEGA.)

Exploring Mars

1650–2050

We are nearing the end of our account. Today, Mars is regularly visited by small machines launched from its neighbour in space, the Earth. They are triumphs of ingenuity and creativity. Nevertheless, some have crashed, or burned, or have been lost in space without accomplishing their missions. In spite of everything, others have prevailed, and have recounted to us the history of Mars as told in this book. More Mars missions will follow, and this history will be re-written; and one day, human beings will assemble the resources to undertake the perilous expedition themselves.

The Viking 2 robot lander, sitting on the plain of Utopia since 3 September 1976, is visited by astronauts after decades of solitude. (Artist's impression courtesy of NASA and Pat Rawlings.)

1 Telescopic voyages

Until the 1960s, Mars was observed – and visualised – from Earth

The length of a Martian year was known in antiquity, but only in the seventeenth century could the study of Mars really begin, with the invention of the telescope.

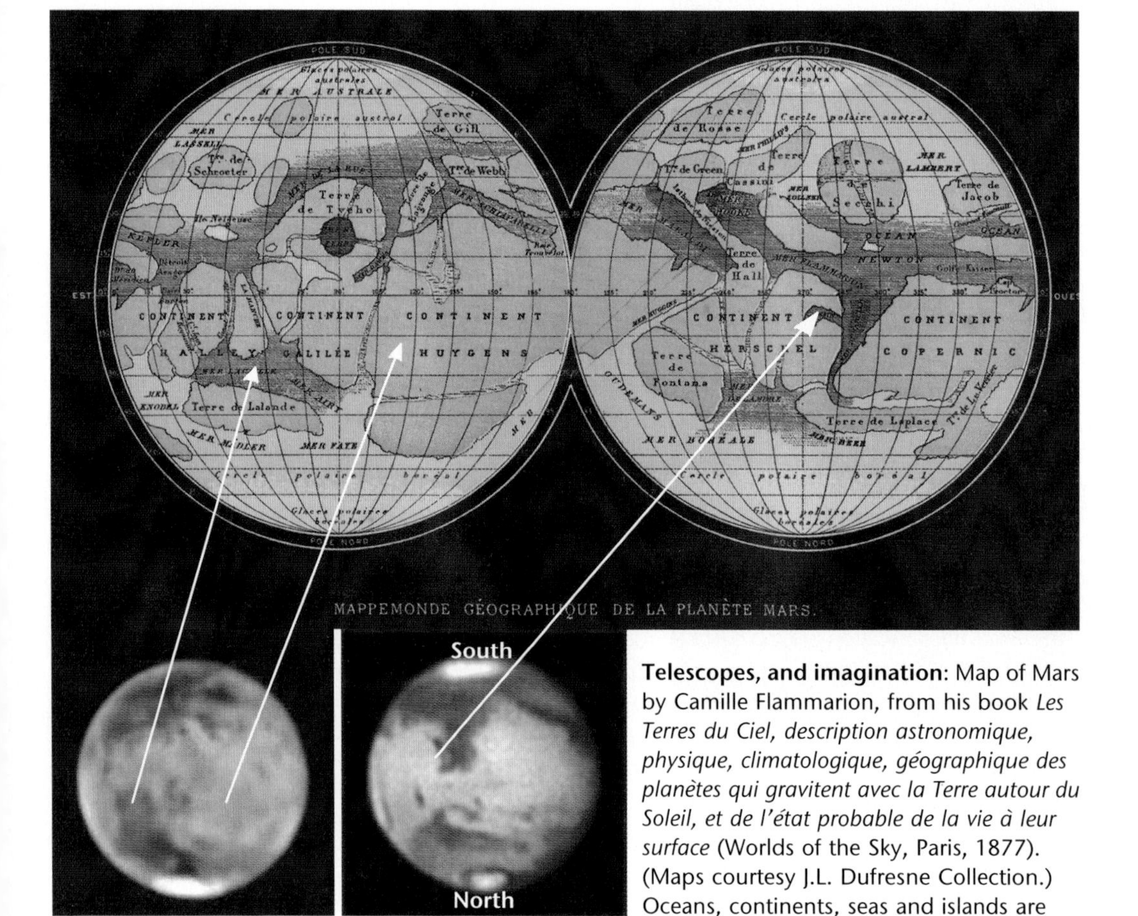

Telescopes, and imagination: Map of Mars by Camille Flammarion, from his book *Les Terres du Ciel, description astronomique, physique, climatologique, géographique des planètes qui gravitent avec la Terre autour du Soleil, et de l'état probable de la vie à leur surface* (Worlds of the Sky, Paris, 1877). (Maps courtesy J.L. Dufresne Collection.) Oceans, continents, seas and islands are charted. On the map, the longitude 0° reference meridian is the same as today's; but, following astronomical tradition, north is downward. The nomenclature proposed by Flammarion has been abandoned. From 1877 onwards, charts prepared by Schiaparelli were more often cited, and his mythologically inspired Latin names (Hellas, Argyre, Syrtis Major etc.) were taken up by Lowell and Antoniadi, to become the basis of modern maps. (Images of Mars below maps courtesy J. Rosenqvist.)

Earth-based telescopic observations are still useful nowadays to monitor the planet between space missions or to test new kinds of instrument. For example, with the aid of a network of radiotelescopes such as the IRAM interferometer (above) on the Plateau de Bure, in France, Martian wind speeds can be measured using the Doppler effect of the microwave radiation emitted by the Martian atmosphere. (Courtesy IRAM, Rambaud.)

Christiaan Huygens made early sketches of Mars in 1659, and, later, Giovanni Domenico Cassini mapped the planet, discovered its polar caps and determined the length of the Martian day (24 hours and 40 minutes). Many scientists continued to observe Mars, but it was only in 1777 that real advances were made, thanks to telescopes manufactured by William Herschel, a musician and organist who became one of the most successful astronomers of his time. He is best known for his discovery of the planet Uranus in 1781. Herschel measured the axial tilt (obliquity) of Mars, and his observation of the growth and shrinking of the polar caps led him to conclude that Mars has seasons like those of Earth. The presence of clouds and the way in which stars were occulted by the edge of the Martian disc suggested to Herschel that the planet had an atmosphere. In the nineteenth century, with better optical instruments at their disposal, astronomers began charting the planet in detail and allotting names to its dark and light areas, thought to be oceans and continents.

1877 became a 'red-letter' year. Mars was then relatively close to the Earth, and the American astronomer Asaph Hall discovered the two satellites, Phobos and Deimos. Giovanni Virginio Schiaparelli, director of the Milan Observatory, created charts which rival in quality more modern versions. He named many features, and most of these names are still used. His best remembered observations were of rectilinear features which he called *canali*. Poor translations converted his channels into 'canals' in English and 'canaux' in French, and the idea of their artificial origin was born. This concept caught the imagination of a brilliant and rich American, Percival Lowell, who, in 1896, had an observatory built in the mountains of Arizona. He developed a picturesque theory, which he maintained until his death in 1916: the canals had been engineered by an intelligent race, threatened by drought. As recently as 1962, discussions on the origin of the canals were still appearing in scientific journals.

Little by little, the study of Mars proceeded, with the aid of modern astrophysical tools such as spectroscopes, polariscopes and radiometers. Since 1896, it had been known that Mars has no oceans. Henceforth, the dark patches would be ascribed to vegetation rather than water. From 1920 onwards, studies focussed on the Martian atmosphere: scientists concluded that it contained very little oxygen and very little water: only carbon dioxide had been detected (in 1947). The density of the atmosphere remained a matter of debate until the coming of the space age.

Earth date	Season on Mars (solar long.)	Min. distance Earth-Mars AU (1)	App. diameter Mars (2)
28 Aug 2003	S. spring (250°)	0.373	25.1″
7 Nov 2005	S. summer (320°)	0.470	19.9″
24 Dec 2007	N. spring (07°)	0.600	15.8″
29 Jan 2010	N. spring (45°)	0.664	14.1″
3 Mar 2012	N. spring (78°)	0.674	13.9″
8 Apr 2014	N. summer (113°)	0.621	15.1″
22 May 2016	End N. summer (157°)	0.509	18.4″
27 Jul 2018	S. spring (220°)	0.386	24.2″
13 Oct 2020	S. summer (296°)	0.419	22.3″

(1) One Astronomical Unit (AU) is the mean distance between the Sun and the Earth (149.6 million km).
(2) The apparent diameter of Mars as seen from Earth is an angle expressed in arc seconds. For comparison, the apparent diameter of the Moon as seen from Earth is of the order of 2000 arc seconds (half a degree). Mars appears to be a hundred times smaller.

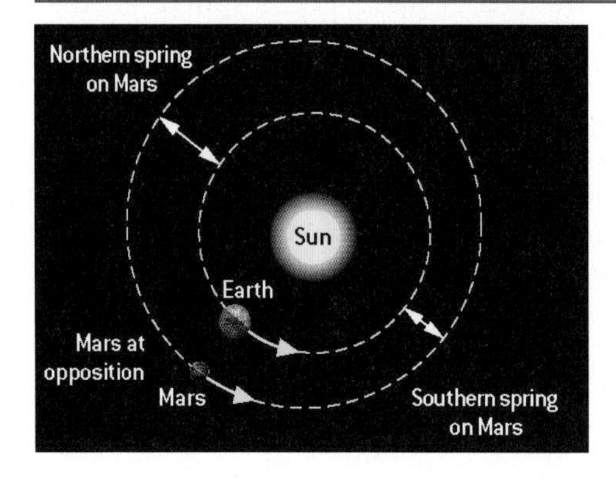

When is the best time to observe Mars from Earth? In its orbit, Mars reaches its nearest point to the Earth about every 26 months. This is known as **opposition**. At this time, Mars is opposite the Sun in the sky as seen from Earth. Mars is best observed only during the two months preceding and following opposition. Since Mars' orbit is eccentric, its apparent size at opposition will vary, and the planet may look twice as big in some years as it does in others (see table). The most favourable oppositions (in red above) occur approximately every 15 Earth years.

2 First visitors: the Mariner probes

NASA explores Mars: 1964-1972

The effective exploration of Mars began on 14 July 1965, when Mariner 4, launched in November 1964, passed over the planet at an altitude of 9,850 kilometres. Mariner 4 was not the first space vehicle to approach Mars. The Soviet space probe Mars 1 had already flown by the planet, but Mariner 4

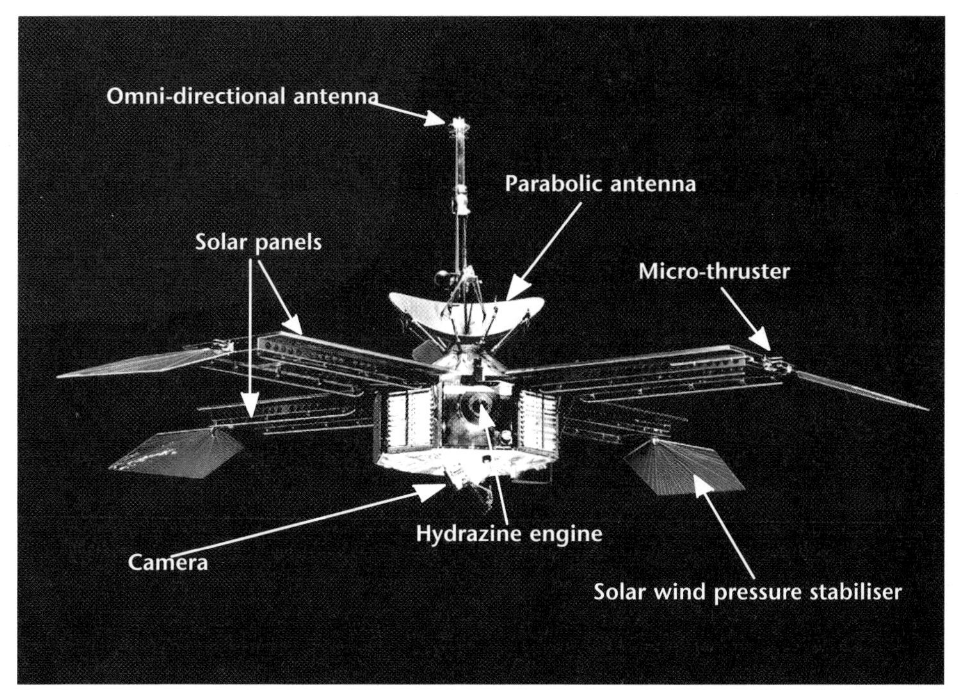

Leading the way: Mariner 4. Launched on 28 November 1964, after the failure of its twin Mariner 3, Mariner 4 weighed 260 kg. Its energy was drawn from solar panels, and it carried a primitive computer, gyroscopes and an attitude and orientation control system with thrusters. Communication with Earth was via a parabolic antenna and a smaller, omnidirectional antenna. The velocity of the probe relative to the Earth was known with extreme accuracy, thanks to measurements of the Doppler shift of an ultra-stable radio signal. Its co-ordinates in interplanetary space were worked out using a model, and in-flight trajectory corrections were made with the aid of a small rocket engine fuelled by hydrazine. Even at this early stage, Mariner 4 carried all the types of system which would assure the success of future NASA missions to Mars. (Courtesy NASA and Dave Williams.)

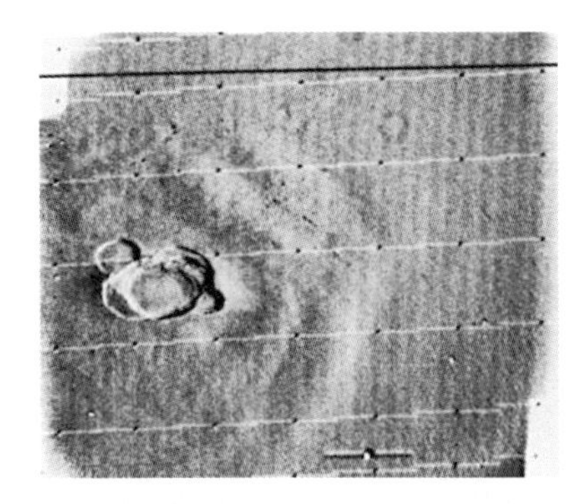

Mariner 9 discovers Olympus Mons looming above the dust, on 27 November 1971. (Courtesy NASA.)

transmitted the first images to Earth. The NASA craft carried a camera, and sent 22 photographs of ancient terrains in Mars' southern hemisphere. Disappointment reigned: no canals, no Martians, but craters by the dozen... Mars looked as dead as the Moon! Moreover, indirect analysis of the nature of the atmosphere indicated a surface pressure of less than 7 millibars, much more tenuous than expected in even the most pessimistic forecasts.

Four years later, in 1969, just a few days after the successful Apollo 11 mission to the Moon, Mariner 6 and Mariner 7 sent back

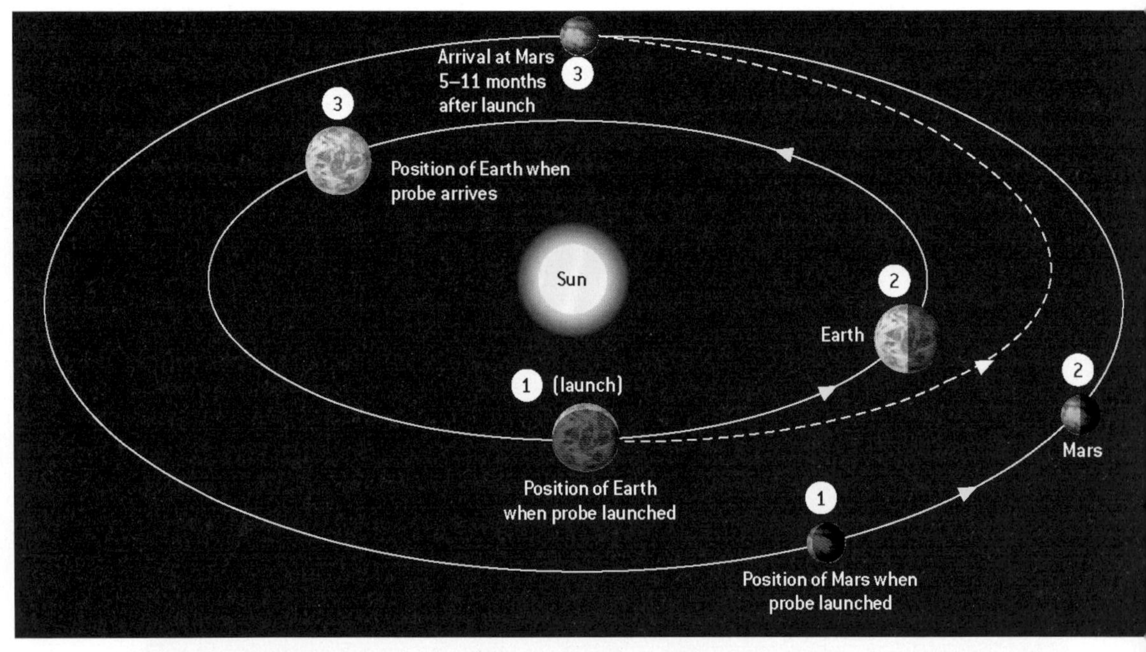

How can we reach Mars? To reach Mars orbit, spacecraft are launched from Earth into a solar orbit, of which the aphelion (the point furthest from the Sun) is close to the orbit of Mars. This strategy minimises fuel consumption. After the initial thrust, the motors are no longer used, except for slight corrections of the trajectory. Since Mars and the Earth are both moving in their orbits at different speeds, the launch must take place during a precise 'launch window' of about 3 to 4 months before opposition (when the Earth is at its closest to Mars, 'overtaking' it in its journey around the Sun). Like oppositions, launch windows occur every 26 months and last only a few weeks. Depending on the year of launch and the rocket used, a flight to Mars will take between 5 and 11 months. The Mariner 4, 6 and 7 probes were restricted to fly-bys of Mars, continuing on into orbits around the Sun. However, subsequent probes were able to 'brake' using retro-rockets, inserting themselves into an orbit around Mars, or landing upon its surface, decelerating ('aerobraking') as they flew through its atmosphere.

two hundred photos of Mars, and made the first observations of the surface and the atmosphere in the infrared and ultraviolet. Unfortunately, these two probes were passing above the same cratered and relatively flat terrain which Mariner 4 had photographed. Mars still seemed a dead world.

However, in 1971-1972, interest in Mars was reawakened when NASA used powerful launch vehicles designed to place twin probes, Mariner 8 and Mariner 9, weighing 980 kg each, in orbit around the planet. Mariner 8 was launched on 8 May 1971, but fell into the ocean after the failure of the Centaur rocket. Mariner 9 was more successful. Launched three weeks later, it was to become the first artificial satellite of a planet other than the Earth. Mariner 9's harvest of data would revolutionise our understanding of the red planet, and would finally reveal its true, more active nature.

When Mariner 9 arrived at Mars on 14 November 1971, the planet was in the grip of the greatest dust storm ever observed. The whole of its surface was obscured. Once the dust had settled, the probe carried out a complete photographic survey of the Martian surface, consisting of more than seven thousand images. A new planet met our eyes: volcanoes, canyons, river beds, spreading glaciers... During the northern winter and for part of the spring, Mariner 9 examined the surface, the atmosphere, the polar caps and the clouds (in the thermal infrared, ultraviolet etc.). More than 30 years on, the data from this spacecraft are still used in scientific publications.

3 Soviet setbacks

The first to Mars: but the Soviets were to know many failures, and rare (if partial) successes

On 10 October 1960, scarcely three years after the launch of the first artificial Earth satellite, Sputnik 1, the Soviet Union sent the first Mars probe into space: Marsnik 1 was to fly by Mars on a photographic mission. However, it never reached Earth orbit. Eleven years went by, and eight more missions had failed when, at last, the Soviet probes Mars 2 and Mars 3 sent their first photographs of Mars, in December 1971. Meanwhile, the USA had already 'conquered' Mars with their Mariners 4, 6 and 7, and Mariner 9 had crowned the achievement with its successes, also in 1971. It was a bitter pill for the Soviets when the pre-programmed cameras of Mars 2 and 3 vainly photographed a planet entirely cloaked in the great dust storm of 1971. The few images and sparse data gleaned by the Soviet probes did not bear comparison with the impressive harvest of Mariner 9 (see pages 187–189).

In 1973, the Soviet Union took advantage of the two-year delay in the Viking programme to 'score double points': the launching of four space vehicles in July and August. Two of them were orbiters (Mars 4 and 5) and two were landers (Mars 6 and 7). Mars 4 and 5, weighing in at about 3500 kg, were equipped with a battery of scientific instruments designed to observe Mars 'from every angle' (i.e. at all wavelengths). The landers, Mars 6 and 7, at just over 600 kg each, carried an array of meteorological detectors and a mass spectrometer for measuring the chemical composition of the atmosphere. Great revelations were expected of the Mars quartet.

Sadly, the gremlins were poised to strike. A few weeks before the launches, flaws were identified in the manufacturing process of microchips in the craft. It was too late to remedy this, and the decision to launch went ahead. Mars 4 and Mars 7 went off course and were lost in space. Mars 6 actually landed, but lost contact with Earth. Even data transmitted during descent proved unusable. Finally, Mars 5 managed to achieve Mars orbit, but ceased transmitting after 22 orbits.

This catalogue of failures led to an unalterable decision: the Soviets abandoned their Mars effort and concentrated instead on Venus. The decision bore fruit: to this day, the Venera missions are the only ones to have touched down on Venus, and subsequently photographed and analysed its surface. More Soviet probes were sent, to Phobos in 1988 and to Mars in 1996. Two of them failed (Phobos 1 and Mars 96), and one was a partial success (Phobos 2). Was Mars, the red planet, displaying distinctly anti-Soviet tendencies?

1. Mars 1, the first human-made object to approach Mars (without radio communication), on 19 June 1963.

2. Mars 2 (above) and Mars 3 (1971), orbiters and descent modules.

3. Mars 5, 6 and 7. This Cuban stamp commemorates "Mars 5, the satellite of Mars", but the image probably shows Mars 6 or 7, carrying the landing module.

4. Phobos 2 (1989) – a partial success.

Much of the content of this book is based on NASA discoveries, and there is relatively little mention of the enormous Soviet Mars exploration programme. A score of missions, more than 60 tons of equipment aimed at Mars – and almost all to no avail. All there is to show for all this effort are a few scientific results; until recently, because of the Cold War, little was known about the missions themselves. Hence, the stamps of Communist countries were for a long time one of the rare sources of information on the extraordinary Soviet projects. (Stamps from the collection of planetologist N. Mangold).

4 Viking

A new American challenge: landing on Mars

Night launch of Viking 2 from Cape Canaveral, Florida, on 9 September 1975, using the powerful Titan 3/Centaur launch vehicle. (Courtesy NASA.)

Emboldened by the success of Mariner 9 in 1971, NASA set itself the goal of placing a module on the surface of Mars. Engineers at the Jet Propulsion Laboratory (JPL) were conscious that, although the Soviets (see pages 190–191) had been successful with their missions to Venus, an apparently far more inhospitable planet, their Mars effort had seen far fewer rewards. With this in mind, NASA's preparations for their Viking missions were exhaustive: two identical spacecraft, weighing 3527 kg each, were launched on 20 August and 9 September 1975. They consisted of a Mariner-type orbiter, coupled with a lander weighing more than a tonne. This would be one of the most complex and costly explorations ever undertaken by NASA. The precautions were worthwhile: the 10-month journeys to Mars went without a hitch. Viking 1 entered Mars orbit on 19 June 1976. Its instruments (cameras, infrared radiometer, water vapour detector) were deployed, searching for an ideal landing site: 'not too high, not too windy, not too sloping, not too rocky, not too dusty, not too near

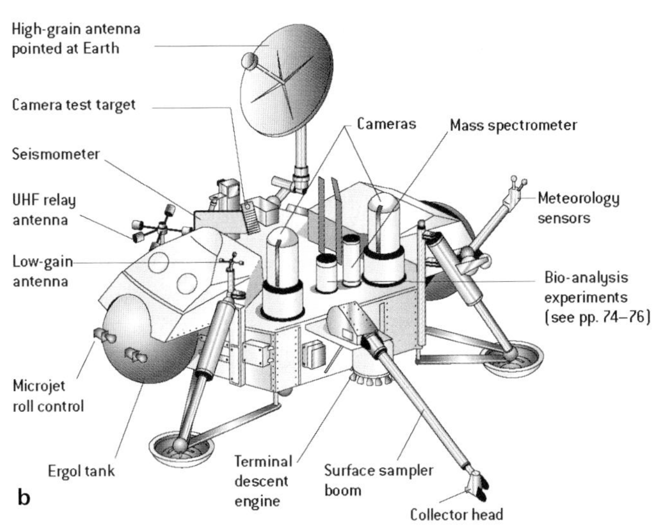

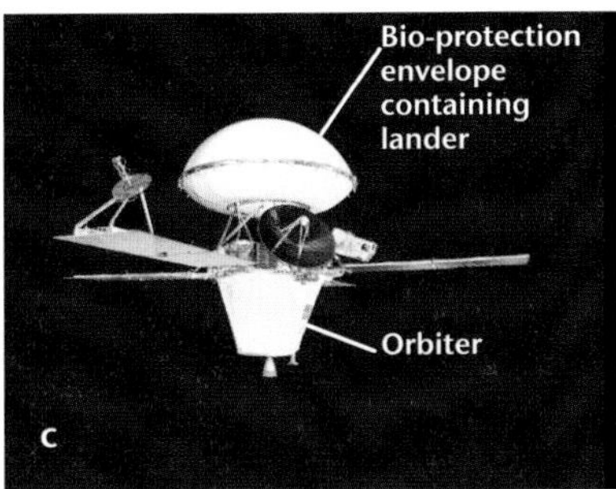

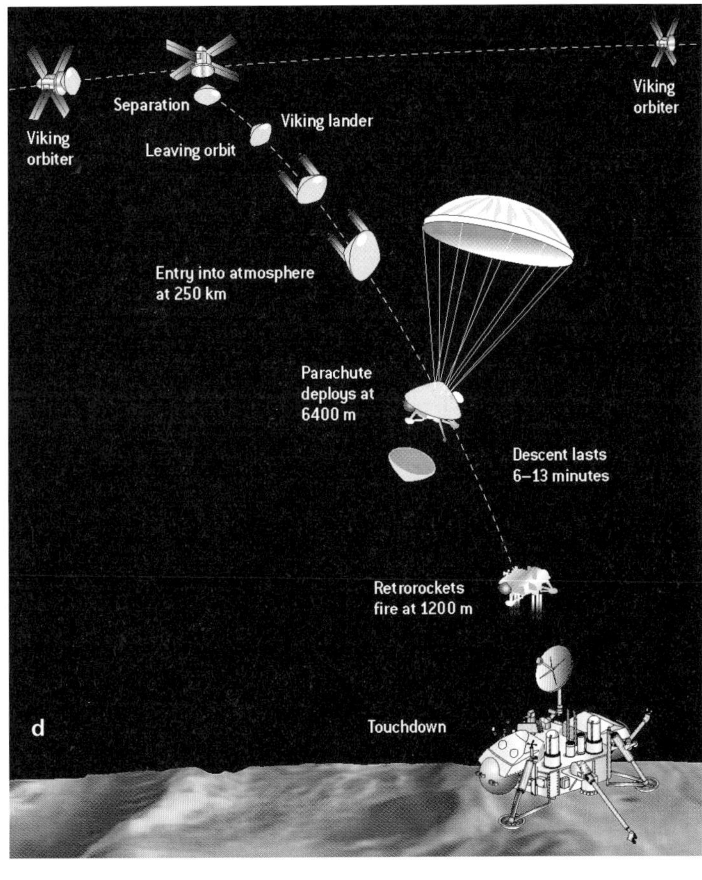

The Viking 1 lander being assembled in 1975 (a), and its components (b). The Viking 1's lander atop its orbiter (c). Before landing, Viking passed through the Martian atmosphere at more than 4 km/s, protected and decelerated by its heat shields. At 6400 metres, a giant 18-metre-wide parachute, suitable for Mars' thin atmosphere, was deployed. As happened earlier with the lunar modules, retro-thrusters slowed the lander during the last few hundred metres of its descent, auto-piloted by the onboard radar (d). Before leaving Earth, the two probes had been carefully sterilised, so that Mars should not be contaminated, in accordance with an international agreement signed in 1968.

On the surface, a plutonium-based radioactive source provided energy for the module. A parabolic antenna was used for communications with Earth, either direct, or relayed *via* the orbiters. As well as cameras and the necessary equipment for biological experiments, the Viking landers also had meteorological instruments, a mass spectrometer and an X-ray fluorescence spectrometer for chemical analysis, and a seismometer. (Courtesy NASA.)

the pole'. It was almost impossible to find such a site from orbit, and NASA had to delay the landing (originally planned to coincide with the American Bicentennial on 4 July; national pride was eventually satisfied, however, in 1997, when Mars Pathfinder landed on that date). Viking 1 finally touched down in triumph on 20 July (the eve of Belgium's National Day!), on Chryse Planitia, at latitude 22° N. It just missed, by a few metres, a large boulder which could have ended the mission there and then. Images of Mars as seen from its surface became front-page news, but the primary object of the mission was the search for life. As the days went by, the scientists overseeing the Viking experiments (see pages 83–85) pieced together their conclusions: officially, 'no life', but unanimity was lacking. Like its sister ship, Viking 2 landed unhindered, this time on the other side of the planet on the plain of Utopia, at latitude 48° N. The results of its experiments were the same. Nevertheless, the Viking mission was counted a success: the expected working lifetime of the modules was 90 days, but they continued to render loyal service for more than 6 years. The last contact, with Viking 1, took place on 13 November 1982. The trail had been blazed for further orbital and surface-based studies of the nature of Mars. Viking may not have provided as revolutionary an advance in our understanding of Mars as Mariner 9, but suffice it to say that we owe to the Viking missions much of the data used in this book.

5 A difficult exploration

Space probes so complex that the smallest error could prove fatal

Rarely has there been a goal so difficult to attain. A score of Soviet missions had registered only four partial successes. The last, Mars 96, was launched in 1996. It was a very ambitious spacecraft of more then 6 tonnes, including four landers. However, the mission was lost in Earth orbit, shortly after launch, and fell into the Pacific not far from the Bolivian coast. The nine successful American missions cannot hide the fact of their five major failures. 1999 was indeed an *annus horribilis* for NASA, with the loss, in just a few weeks, of Mars Climate Orbiter and Mars Polar Lander, with its two Deep Space 2 soil-penetrating microprobes. Launched in the same year, the Japanese Nozomi probe suffered a similar fate after a prolonged itinerary around the Sun. Nozomi's troubles began in 1998, when a malfunctioning valve in the propulsion system prevented its injection into a trajectory towards Mars. Japanese mission controllers cleverly devised a new and much longer flight plan, which would have brought Nozomi to the vicinity of Mars in late 2003, but a large solar flare put paid to the mission in April 2002, damaging its electrical and heating systems. With its fuel frozen, the spacecraft became uncontrollable, and was declared lost in December 2003.

Nozomi's misadventure was just one among the many: there is, to date, a whole spectrum of possibilities for accidents, among them seven failures at the always delicate moment of launching. Direct human error has played its part: the Soviet probe Kosmos 419 burned up in Earth orbit in 1971, one of its controllers having programmed insertion from Earth orbit into a Mars trajectory for 1.5 years instead of the planned 1.5 hours! Mars Climate Orbiter disintegrated when it was put into orbit at an altitude of 57 kilometres instead of the correct 120 kilometres, because one team used English units while the other used metric units for trajectory correction instructions. Sometimes, a radio blackout means that the causes of failure are unknown, as happened in the case of Mars Observer, which was lost just before it entered Mars orbit in 1993. Similarly, Mars Polar Lander and Beagle 2 both disappeared without trace as they descended towards the Martian surface, in 1999 and 2003 respectively.

Sometimes, discoveries may be made because things go wrong. A good example of this is the observation of magnetic anomalies in the Martian crust by the magnetometer aboard Mars Global Surveyor (MGS) as the probe passed repeatedly over the surface at altitudes of less than 120 kilometres. The magnetometer had originally been designed to go aboard Mars Observer, a large vehicle which was never meant to descend below 350 kilometres: for this reason,

'Mars shots'. The dates refer to launches.

	USSR (or CIS)	USA	Others
Launch failure	Marsnik 1 (1960) Marsnik 2 (1960) Sputnik 22 (1962) Sputnik 24 (1962) Mars 1969A (1969) Mars 1969B (1969)	Mariner 8 (1971)	
Lost *en route*	Mars 1 (1962) Zond 2 (1965) Zond 3 (1965) Kosmos 419 (1971) Phobos 1 (1988) Mars 96 (1996)	Mariner 3 (1964)	Nozomi (Japan, 1998)
Failed to enter Mars orbit	Mars 4 (1973)	Mars Observer (1992) Mars Climate Orbiter (1998)	
Crash landed	Mars 3 (1971) Mars 7 (1973)	Mars Polar Lander + Deep Space 2a, 2b (1998)	Beagle 2 (UK-Europe 2003)
Partial success	Mars 2 (1971) Mars 5 (1973) Mars 6 (1973) Phobos 2 (1988)		
Total success		Mariner 4 (1964) Mariner 6 (1969) Mariner 7 (1969) Mariner 9 (1971) Viking 1 (1975) Viking 2 (1975) Pathfinder (1996) Mars Global Surveyor (1996) Mars Odyssey (2001) Mars Exploration Rovers (2003) Mars Reconnaissance Orbiter (2005)	Mars Express (Europe, 2003)

it certainly would not have been able to detect the anomalies. After Mars Observer was lost in 1993, a spare magnetometer was fitted to MGS, a smaller probe which would orbit Mars at very low altitudes in order to take advantage of upper atmospheric resistance (aerobraking): the magnetic fields were easily detected. This was not, however, the end of the story. As a result of damage to one of the solar panel roots, MGS was forced into a progressive braking manoeuvre and the number of low-altitude passes was increased, ensuring a thorough cartographic survey of the anomalies.

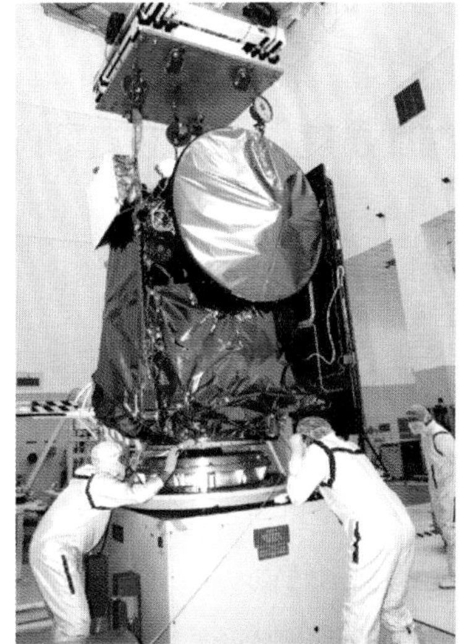

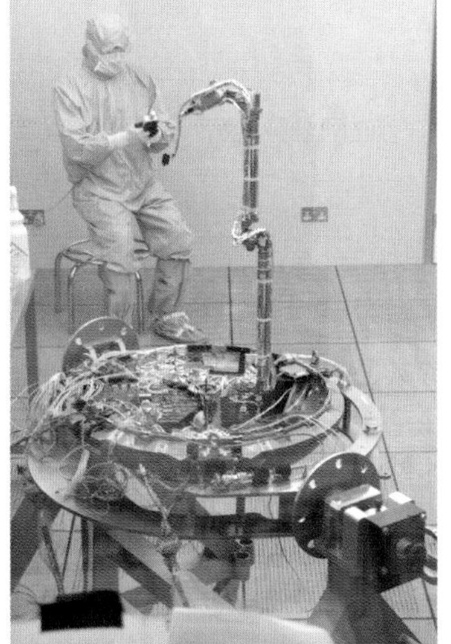

The last 'victims' of Mars: hundreds of thousands of hours of work gone for nothing. (a): Mars Climate Orbiter (1998) (Courtesy NASA, JPL.); (b): Beagle 2 (2003) (Courtesy Beagle 2 all rights reserved.); (c): Mars 96 (1996) (Courtesy IPGP/DT, INSU); (d): Mars Observer (1992) Image kindly supplied by B. Harris); (e): Mars Polar Lander (1998). (Courtesy NASA, JPL.)

6 New American initiatives

Every two years, one or more American probes set out for Mars

After the success of the two Viking missions (1976), seventeen years elapsed before another American space vehicle, Mars Observer, set out for the red planet in September 1992. During the intervening period, NASA had put Mars on its 'back burner' and its goals were elsewhere: Venus (with Pioneer Venus, 1978, and Magellan, 1989), and Jupiter and the other giants (Voyager, 1977, and Galileo,

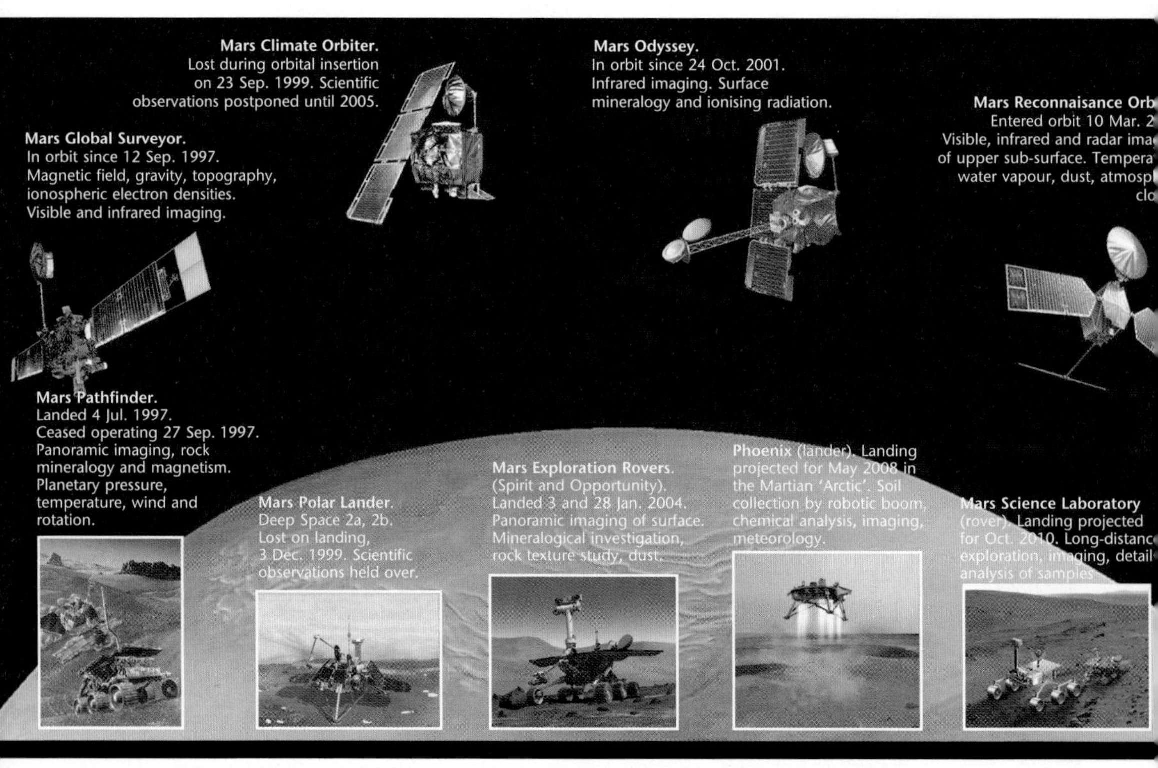

14 years of NASA missions (1996-2010). No fewer than ten American rockets will have been launched toward Mars during this period. The most recent arrival was Mars Reconnaissance Orbiter, carrying a new ultra-high-resolution (30-60 cm/pixel) camera (HiRISE), a wide-angle colour camera (MARCI) and an imaging spectromete (CRISM) similar to OMEGA on Mars Express (see pages 205–207), but with a spatial resolution at least ten time greater (20-30 m/pixel). Another instrument, Mars Climate Sounder (MCS), studies temperatures, water, cloud and aerosols in the atmosphere. The SHARAD radar sounds out the structure of the sub-surface, to a lesser dept than the MARSIS radar on Mars Express (see pages 205–207), but with greater vertical resolution. (Courtesy NASA, JPL. Illustration of Phoenix by Corby Waste.)

Future flights to Mars. (a) Phoenix, which was launched successfully in August 2007, will land in May 2008 in the high latitudes of the northern hemisphere to study *in situ* one of the regions where the Mars Odyssey probe discovered an ice-rich layer beneath a few centimetres of sediments (see pages 173–175). The lander will deploy a robot arm to dig into the soil and bring in samples, which will perhaps contain ice, for analysis in micro-labs. (Courtesy NASA, JPL.)
(b) Phoenix being assembled at the Lockheed-Martin facility in Colorado. The probe is seen base upwards. Its legs have not yet been attached. The name 'Phoenix' was chosen as a reminder that the probe has been constructed from elements initially developed for the (abandoned) Mars Lander 2001 mission, and from the 'ashes' of Mars Polar Lander, lost in 1999. (Courtesy NASA, JPL, Lockheed Martin.)
(c) Mars Science Laboratory (to be launched in late 2009): an enormous 600-kg rover (the exploring rovers already on the planet weigh only 185 kg). Equipped with a radioactive energy source to ensure a long-lasting mission, covering several tens of kilometres, MSL will carry 60 kg of scientific instruments. It will have a set of cameras, from zoom to microscopic, spectrometers to analyse rocks, meteorological instruments, and the SAM micro-laboratory capable of identifying many different mineral and organic molecules by their isotopic signatures (see pages 74–76). The image shows ChemCam, a new kind of instrument which will use a powerful laser to analyse rocks ten metres away, vaporising matter and studying the results with a spectrometer. (Courtesy French Space Agency (CNES) and Los Alamos Laboratories.)

1989). Other events that slowed the pace of planetary exploration were the end of the Cold War, and the Challenger disaster of 1986.

The Mars Observer mission was nevertheless an ambitious one. The probe, which weighed 2487 kg, carried a high-resolution camera (the MOC), an infrared spectrometer (TES) and a gamma-ray spectrometer (GRS), all destined to participate in a mineralogical survey of the surface. Other equipment on Mars

Observer included the MAG magnetometer, the MOLA laser altimeter, and an infrared detector (PMIRR) specially designed for atmospheric studies. The French contribution was a device intended to communicate with a balloon which was to fly over Mars two years later (see pages 211–213). Unfortunately, on 20 August 1993, as the propellant tanks were being pressurised, the probe inexplicably disappeared from the screens, its transmission system having previously been deactivated.

NASA, shaken to the core by the failure of its billion-dollar mission, changed its way of thinking, giving priority to shorter and less expensive projects, which would be more frequent. The Pathfinder (see pages 114–116) and Mars Global Surveyor probes were the first products of this new approach: MGS weighed only 1100 kg, and carried six instruments similar to those on Mars Observer.

The two probes, launched in 1996, would achieve and even surpass their objectives: indeed, MGS functioned for more than ten years, contact being lost on 2 November 2006. Amid the euphoria, a new programme was elaborated: as each 'launch window' came round, every twenty-six months, an orbiter and a small lander would be sent to Mars. The first of these were Mars Climate Orbiter (MCO) and Mars Polar Lander (MPL) in 1998. Moreover, NASA decided that, from 2003 onwards, the exploration of Mars would be complemented by an ambitious project involving the return to Earth of samples from the planet (see pages 214–216), in collaboration with France and Italy.

Sadly, everything fell apart in 1999. All four probes which had been launched the year before (MCO, MPL and the two soil-penetrating microprobes) failed one after the other. Again, the Mars programme was in trouble. NASA postponed its sample return project until after 2015–2020, in order to concentrate on a more progressive programme of exploration. The momentum was not lost, however, and after Mars Odyssey went into orbit around the planet in October 2001, the great success of the two Mars rovers, set down on its surface in 2004, restored NASA's prestige (see next page). We now await discoveries by the 'super-satellite' Mars Reconnaissance Orbiter (which entered Mars orbit on 10 March 2006), by the Phoenix lander (2007) and by Mars Science Laboratory (2009).

7 Spirit and Opportunity: wheels on Mars

The Mars rovers

After the failure of the Mars Climate Orbiter and Mars Polar Lander missions in late 1999, NASA's exploration programme found itself at an impasse. A lander due to arrive in 2001 was cancelled. What NASA needed for the next window of opportunity, in June 2003, was a project both 'safe' and spectacular. In July 2000, a solution emerged: the Mars rovers. The proven Pathfinder landing system was to be used, but now, instead of a fixed station, a roving vehicle capable of travelling several kilometres was to explore Mars, carrying some of the scientific equipment prepared for the 2001 lander. To improve the chances of success, two identical rovers were to be sent to Mars atop different launch vehicles. The problem was that NASA's Jet Propulsion Laboratory had only three years to bring

Self-portrait. The shadow of Opportunity on the flank of Endurance Crater, caught by its navigation camera, 26 July 2004 (sol 180). (Courtesy NASA, JPL-Caltech, Cornell.)

Very complex machines. Each Mars exploration rover weights 185 kg and carries 5 kg of scientific instruments: veritable robot geologists, capable of 'seeing' and of analysing the surfaces of the rocks they encounter. They can even probe the interiors of these rocks, thanks to the Rock Abrasion Tool (RAT). (Courtesy NASA, JPL.)

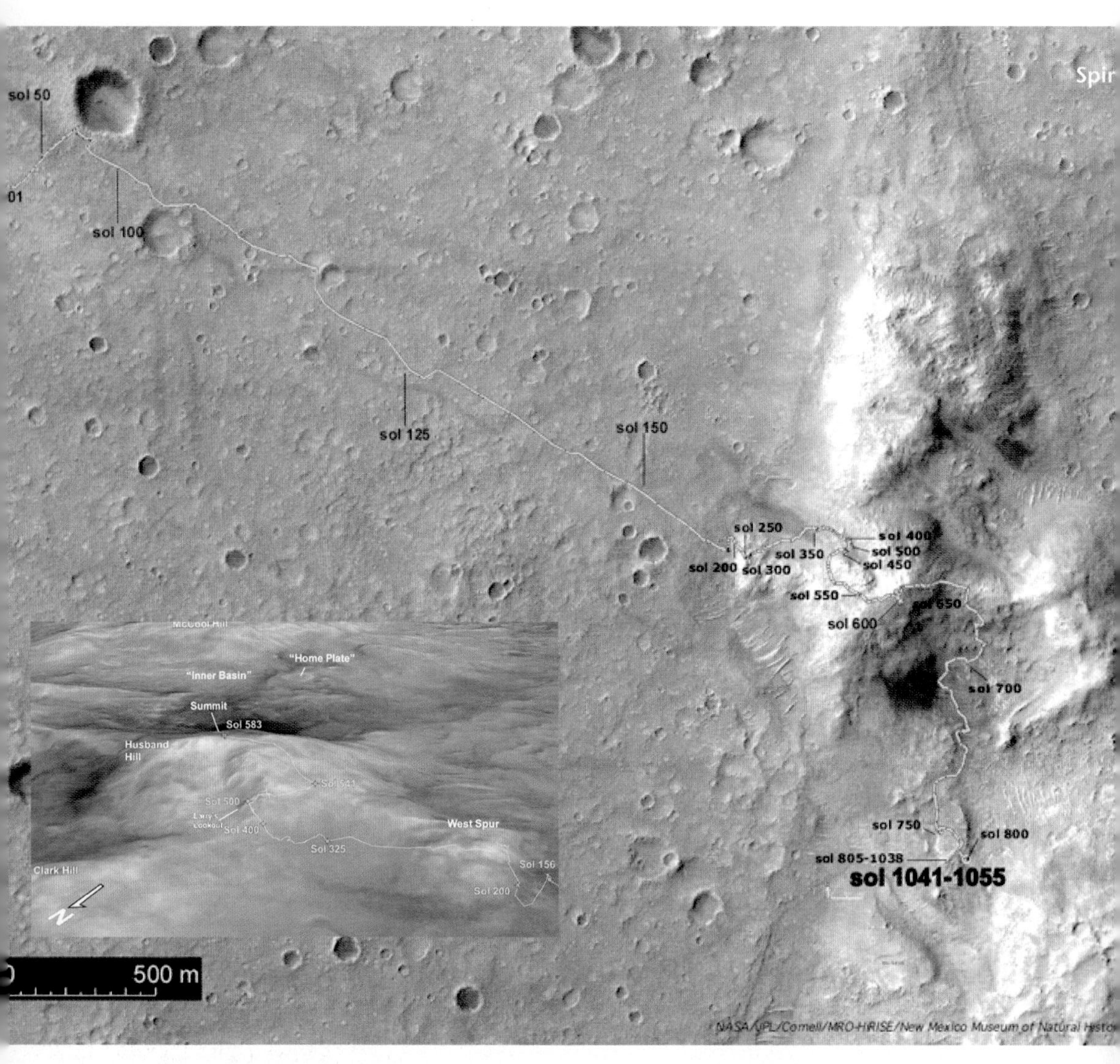

On the trail of Spirit. The journey made by the Spirit rover, day by day, after landing (1 sol = 1 Martian day). After months of driving across the flat basaltic plain of Gusev Crater, Spirit successfully climbed into the Columbia Hills. (Courtesy NASA, JPL-Caltech, Malin Space Science Systems, OSU.)

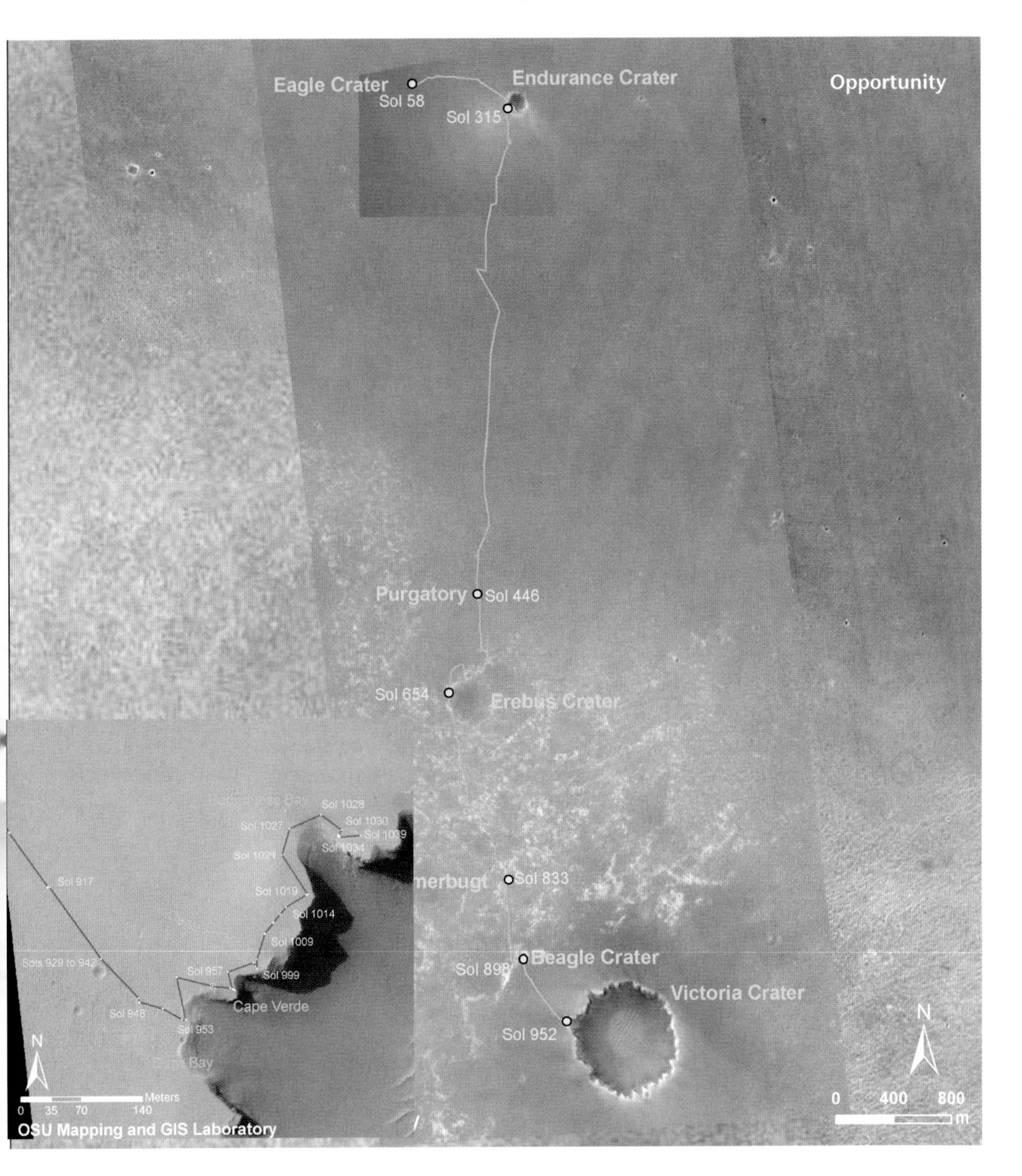

Opportunity traverse map. The names generally refer to craters named after ships and other vessels of exploration. After spending several weeks in Eagle Crater, where it landed, and then several months exploring Endurance Crater (130 m in diameter), Opportunity proceeded southwards to explore Victoria Crater (nearly 800 m in diameter), about 7 km away, at the bottom of the map. (Courtesy NASA, JPL-Caltech, Malin Space Science Systems, OSU.)

to fruition missions of unprecedented complexity. Many difficulties awaited JPL: airbags and parachutes needed to be reconfigured, and a viable design was arrived at only at the last minute. Every test of the rover seemed to throw up yet another fault. It was a very tired team that watched the launches of the two rovers on 10 June and 7 July 2003. Fingers were crossed: would the vehicles even reach their destination? And could they survive there for 90 days, their planned working time on Mars? They succeeded beyond the most optimistic expectations, and at the time of writing, more then three-and-a-half years after they landed, Spirit and Opportunity are still operational, having travelled more than 7 and 11 kilometres, respectively. They have explored the floors of craters, climbed mountains, and, most importantly, have revealed the unexpected geology of two very different sites on Mars, where evidence of the past presence of liquid water is all around (see pages 59–64).

Spirit was the first to land, on 3 January 2004, inside the crater Gusev (see pages 62–64). Everything went smoothly until the eighteenth day, when Spirit fell silent. Its onboard computer seemed to be switching itself on and off, draining the batteries. Several days of worrying and technical wizardry later, the problem was finally overcome, in good time for the team to concentrate on the landing of Opportunity, which occurred without a hitch on 25 January. Luckily, Opportunity came down inside a small crater, where, for the first time, exposed sedimentary bedrocks were visible (see pages 59–61). Thanks to its exceptional location, Opportunity became, for a long time, the researchers' 'pet' rover, but it also had its problems. In the spring of 2005, it was stuck in a small dune for five weeks while Spirit came back into favour by ascending the Columbia Hills, gaining more than 100 metres in altitude – some of it backwards, due to a seized-up wheel! Here, it was to discover increasingly enigmatic kinds of sedimentary rocks.

8 Europe joins in: Mars Express

In 2003, Europe too put a probe into orbit around Mars

The European Space Agency (ESA) has been in existence since the mid-1960s. Early priority was given to astronomical missions and to the study of the Earth's ionised environment, but in the 1980s, Europe began to turn towards the exploration of the solar system. Giotto flew by Comet Halley in 1986, and Huygens successfully landed on Titan, one of the satellites of Saturn, on 14 January 2005. Meanwhile, Mars and the other terrestrial planets, Venus and Mercury, seemed relatively neglected. European planetary scientists had to be content

Mars Express – from dream to reality. Assembling the Mars Express spacecraft in Toulouse. We see the main parabolic dish at the top, and the landing module Beagle 2, which is the black cone near the technician. (Courtesy Beagle 2 all rights reserved.) Most of the seven instruments designed to study Mars from orbit are also seen at bottom right and below the structure:

- The HRSC stereo colour camera, developed by a German team;
- the OMEGA mapping spectrometer, built under the auspices of the Institut d'Astrophysique Spatiale, Orsay. OMEGA is capable of imaging the planet at 352 wavelengths simultaneously in the visible and near-infrared, analysing the composition of the surface and the atmosphere (see pages 59–64);
- the SPICAM ultraviolet and infrared spectrometer, dedicated to the observation of the atmosphere, built by a French team of the Service d'Aéronomie (CNRS);
- the high-resolution PFS spectrometer, an Italian contribution, observing the atmosphere in the near and far infrared;
- the Swedish ASPERA instrument, designed for plasma studies in the space environment encountered by the probe as it orbits Mars (this is mounted separately, and its red cylinder can be seen at the top near the antenna);
- the German MaRS experiment, which performs radio sounding of the Martian atmosphere and ionosphere;
- the Italian/American MARSIS radar, which studies the sub-surface and searches for water (to a depth of 1-2 km).

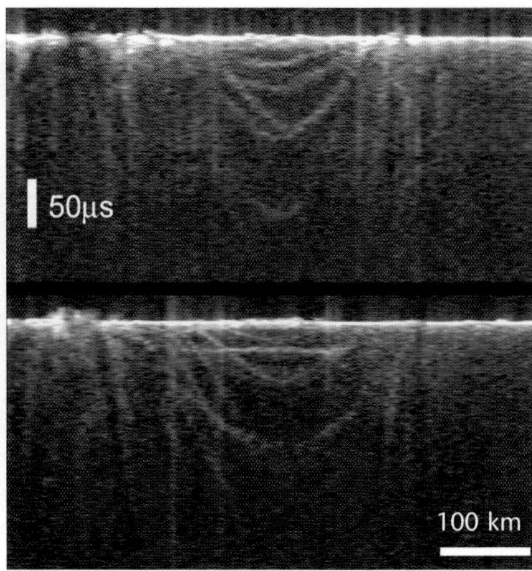

The antennae of Mars Express. The probe sports three long booms, two 20 m long and one 7 m long, which serve as antennae for the MARSIS radar. MARSIS sounds out the sub-surface of the planet, to detect, for example, the presence of ancient, buried craters (shown above as faint white curves, the white line at the top representing the surface itself). The deployment of these antennae was a story in itself: folded accordion-like, the antennae were due to extend when the probe achieved Mars orbit, in April 2004. At the last moment, the operation was cancelled when new calculations showed that the deployment could cause the antennae to strike and possibly damage the spacecraft. After careful computer simulations of the manoeuvre – and much debate – a first section was finally freed one year later, at the beginning of May 2005. The probe encountered no great problems, but the extremity of the antenna seemed not to have been locked onto its final position. New studies were required: the conclusion was that it was necessary to warm up the antennae by exposing them to the Sun, and, on 17 June 2005, the extension of the three booms was finally achieved. (Artist's impression courtesy NASA, JPL, Corby Waste. MARSIS radar images courtesy NASA, JPL, ASI, ESA, Univ. of Rome.)

with 'bolt-ons', collaborating on Mars missions organised by the two superpowers. The former Soviet Union, especially, accepted European instruments on Phobos 1, Phobos 2 and notably on Mars 96. However, the partial failure of Phobos 2 in 1989, the loss of Phobos 1, and the failure of Mars 96 in November 1996 on the launch pad, all served to galvanise minds: European instruments would henceforth be carried on European missions! After the Mars 96 disaster, France and Germany undertook a feasibility study of a small satellite which would carry similar instrumentation to that aboard Mars 96. ESA quickly took this up, and a new kind of mission was developed in record time: Mars Express.

Mars Express was launched on 2 June 2003. On Christmas night of that year, it successfully entered Mars orbit. Scientists, all personalities in their own right, and from all around Europe (French, Italian, German, Swedish etc.) form the Mars Express team, and there are lively debates on mission objectives, observing strategies, the distribution of resources... but everything works fine! Thanks to its state-of-the-art instruments, Mars Express has revolutionised many areas of Mars science.

Beagle 2: the dream that failed. Mars Express carried Beagle 2, a small 30-kg module built by a British team. Beagle 2 was designed to plunge into the Martian atmosphere, land and study the surface material with a battery of miniaturised instruments. The original *Beagle* was the vessel upon which Charles Darwin, father of the evolutionary theory, served his apprenticeship as a naturalist. Sadly, after a perfect 'cast-off', Beagle 2 fell silent, and was assumed to have crashed. The little craft had gained a certain popular following in Britain, where its loss came as a disappointment to many. The exact cause is still unknown. (Courtesy ESA, Medialab.)

9 The future of European Mars exploration

After Mars Express' first steps: the Aurora Programme

Proud of the success they have had with Mars Express, the Europeans want to extend their ambitions: there will be further exploration of Mars within the new Aurora programme, which was decided upon at the end of 2005. The originators of Aurora were conscious that, one day, the study of Mars could go beyond mere scientific research and look to new horizons, with humans involved in exploration. So they brought into the programme researchers already versed in the fields of ESA astronautics and the International Space Station, and to draw on associated resources.

Aurora. A few decades from now, astronauts may come across the frozen robot representative of the first Aurora programme mission, ExoMars. If one of the astronauts is a European, the encounter will symbolically seal the success of the programme, which aims to promote Europe's exploration of Mars by means of robotic probes in the years until 2020. Thereafter, the objective is to participate in an international human exploration of the planet. (Courtesy ESA, P. Carril.)

Another 'descendant' of Mars Express: Venus Express. In 2001, the European Space Agency (ESA) considered the prospect of using the technology already developed for Mars Express to make a similar craft at much less cost. What mission would this craft undertake? Opinions were sought within the scientific community, and one proposal stood out: why not send 'Mars Express Mark 2', carrying almost the same instruments, to Earth's other neighbour, Venus? Venus Express was developed in record time, and successfully launched on 9 November 2005. (Courtesy ESA, C. Carreau.)

ExoMars

The Aurora programme aims to develop the necessary know-how and technologies for the future large-scale exploration of Mars. In 2013, the first stage will be the landing and deployment on Mars of a rover called ExoMars. Its objectives will be both technological and scientific: the search for signs of ancient life, and the investigation of the evolution of the planet's habitability. To this end, the rover will carry a drill capable of going down to a depth of a few metres, to reach rocks and soils unchanged by alteration processes which have worked upon the surface (see pages 83–85). In these samples, exobiologists hope to find chemical clues to primitive Martian life, with analyses taking place in a dedicated mobile laboratory. ExoMars will use a radar device to guide the drill and detect the presence of geological layers and ice.

Another aim of ExoMars will be the study of the history of Mars' habitability, better to understand the kind of environment that future Mars astronauts will be faced with. It is therefore envisaged that the rover could be accompanied to the surface by a miniature geophysical and environmental package, called GEP, attached to the descent module, and operational for several years. This may be the first element in a whole network of such stations on Mars in the future. The planet's current geological activity would be investigated for the first time with state-of-the-art detectors, such as a seismometer able to detect any possible 'Marsquakes'.

So, ExoMars could be the realisation of a long-held ambition: exobiologists, disappointed by Viking's negative results 30 years ago, and geophysicists, still remembering how Mars 96 failed, and how the USA and Europe then shelved Mars projects, now have the Aurora mission to look forward to.

10 Imagination serves exploration

Moles, insects, balloons, 'planes': all on file at the space agencies

Cryobot is designed to melt, explore and analyse the ice of Mars' polar caps. (Courtesy NASA, JPL.)

Geologically, Mars is as diverse as the Earth. In a few years' time, when the gross structure of Mars has been surveyed, the study of this geological diversity will become the principal challenge as the scientific exploration of Mars continues.

Colonies of 'biomorphs'. How can we explore Mars rapidly? Researchers at NASA's Jet Propulsion Laboratory, taking their inspiration from the natural world, imagine a solution to this problem based on colonies of biomorphic robot explorers. These robots might crawl on caterpillar tracks or fly down to the surface like winged seeds, intercommunicating, dispersing to investigate the planet, and re-assembling in force if alerted to a site of scientific interest. (Courtesy NASA, JPL.)

Success will depend on the number and the mobility of the probes employed. They will have to dig, fly, crawl and communicate, in order to explore exhaustively the Martian surface and what lies below it. A number of projects, which would not look out of place in works of science fiction, are on the cards.

Digging into sand and ice

The surface of Mars, exposed to ultraviolet radiation and a chemically aggressive atmosphere, is not an ideal place to detect original traces of ancient sedimentary activity, or find any organic compounds. Digging is essential. For this reason, the European Mars Express probe (2004) carried a 'mole' attached to its lander, Beagle 2, with a view to examining the topmost metre of the surface (see pages 205–207). Weighing less than a kilogramme, the mole was able to crawl across a sandy surface at a rate of 1 centimetre every six seconds, powered by a spring mechanism. With its sample safely gathered, Beagle 2 would then have winched in the 3 metres of cable, recovering the mole and its 'sample', and analysis would have begun. When Beagle was lost, both NASA and ESA started work on a drilling system for post-2010 missions such as ExoMars (see previous pages). This latter mission will give a 'second chance' to the tethered mole technology, which will be used to deploy a suite of sensors aimed at measuring planetary surface heat flow for the first time on Mars.

Now, the polar caps on both Mars and the Earth represent planetary climatic archives. 'Cryobot', a robot designed to melt its way down through tens of metres of ice, is currently being developed at the Jet Propulsion Laboratory (JPL) in California. Its mission would be to analyse the various layers of ice as it encounters them.

Flying and floating in the 'air'

For an aeroplane to be able to fly though the Martian atmosphere, it would need to travel five times faster than through the air of Earth, or have wings commensurately larger. Mars' gravity is 2.7 times weaker than Earth's, and atmospheric lift is about 100 times less. Far from being discouraged, NASA engineers have come up with a number of prototypes. One of them is ARES, a 'plane with a wingspan of about 5 metres, which may one day fly though the skies of Mars at an altitude of 1.5 kilometres, securing detailed images of Valles Marineris or taking profiles of magnetic anomalies at unprecedented resolutions.

Helium balloons are another possibility. Such a balloon, made in France, was to have been carried aboard the Russian Mars 98 mission, but cutbacks in the Russian space programme meant that the mission had to be abandoned.

Proposed Mars Airplane. Such an aircraft will have to be extremely light, given the low density of the Martian atmosphere. It will have to carry an energy source, and possess the ability to deploy in mid-air. (Courtesy NASA, JPL.)

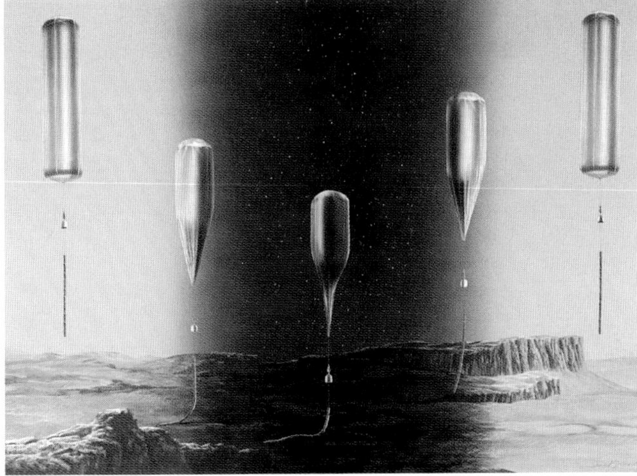

Exploration balloon. The French space agency CNES actually developed such a balloon for the Russian Mars 98 mission, which was cancelled. The balloon held up to 6000 m^3 of helium in a cylindrical container 45 m high and 12 m across, made from mylar film 6 micrometres thick. The whole thing weighed 60 kg (science payload 30 kg, balloon 30 kg). The lack of buoyancy during cold Martian nights (represented by the dark band on the illustration) causes the balloon to drag its serpent-like ballast line ('guide-rope') along the surface. The line would have acted as an antenna, performing radar soundings of sub-surface material. (Courtesy CNES, D. Ducros.)

11 The Grail: returning samples

Without 'fresh' samples, our knowledge of Mars will remain incomplete

The challenge inherent in all the one-way missions to Mars over the last 30 years, and one that will prove much costlier and more technologically demanding, will be the returning to Earth of samples from the red planet. Is it really worth all the

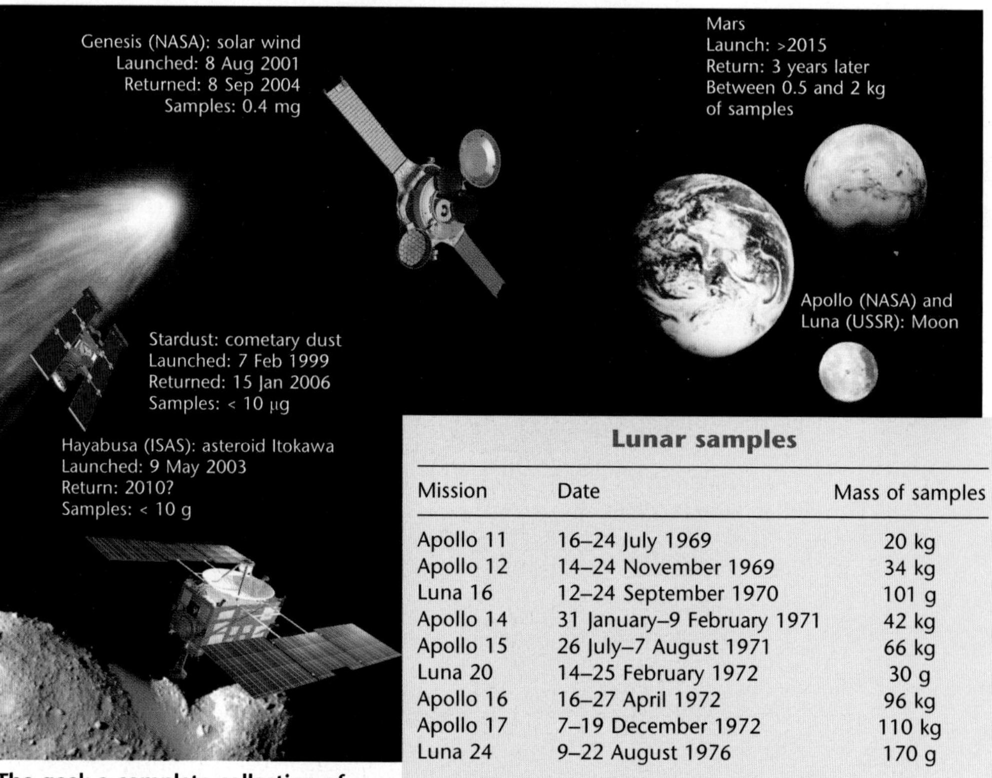

Lunar samples

Mission	Date	Mass of samples
Apollo 11	16–24 July 1969	20 kg
Apollo 12	14–24 November 1969	34 kg
Luna 16	12–24 September 1970	101 g
Apollo 14	31 January–9 February 1971	42 kg
Apollo 15	26 July–7 August 1971	66 kg
Luna 20	14–25 February 1972	30 g
Apollo 16	16–27 April 1972	96 kg
Apollo 17	7–19 December 1972	110 kg
Luna 24	9–22 August 1976	170 g

The goal: a complete collection of samples of the solar system. Having secured lunar material in 1969 and during the 1970s (see table), space agencies have recently considered challenging programmes for returning samples from other solar system bodies. NASA's Genesis mission sampled the solar wind, although its capsule crashed to the ground on return. The Japanese Hayabusa mission (ISAS, Japan) achieved a landing on the asteroid Itokawa, and is now facing a challenging return journey to Earth, after the failure of some spacecraft subsystems. The NASA Stardust mission succeeded in bringing back dust from comet Wild 2, on 15 January 2006. When will the Mars sampler take off? (Courtesy NASA, ISAS, JAXA.)

The return: a technological challenge. Martian gravity is twice as strong as that of the Moon. A launch velocity of 3.5 km/s permits insertion into a low orbit, and an extra 1.5 km/s allows escape into space. One idea is to use a small rocket to put samples into orbit (a), and a second element, a satellite capable of manoeuvring to a rendezvous with the rocket, and then returning the samples to Earth. All these activities will be very demanding of fuel. Will less fuel-hungry techniques, such as electrical propulsion and aerocapture, be employed to initially bring the satellite into orbit? Perhaps the first Martian samples will be limited to atmospheric material and dust, gathered during a rapid fly-by of the planet. Such was the objective of NASA's SCIM mission which was pre-selected for a launch in 2007, but finally not confirmed (b). (Courtesy NASA, Arizona State University.)

Dangerous samples? Martian samples will be examined in a limited number of high-security laboratories, in sterile chambers like the one illustrated, which contains lunar material brought back in 1972 by the Apollo 17 astronauts. Not only does this procedure protect the samples from Earthly contamination, but it also means that they are quarantined for long enough to ensure that (very hypothetical) dangerous Martian life forms are not present. (Courtesy NASA.)

effort? If, as is widely believed, the SNC meteorites are of Martian origin (see pages 30–32), then planetologists already have tens of kilogrammes of Mars rock. What more could we expect to learn from a few more kilos delivered to us by some future sampling missions?

First of all, such 'extra' samples would be taken from the planet's surface, while the SNC meteorites have probably been ejected from deep within the crust. New samples would therefore contain evidence of changes in the surface material due to the action of the atmosphere and water, sedimentation and hydrothermal effects. They might even carry the traces of any ancient life on Mars. Moreover, they will have been protected from any terrestrial contamination.

The fundamental question for scientists is what kind of samples to aim for. What are they trying to investigate? To determine the chronology of Mars, it would be useful to have samples from a site where lava has flowed. At present, the only method of establishing relative chronologies is to count meteoritic impact craters, and this is by no means an exact science. To search for traces of life, a site where sedimentation has occurred, or an ancient river bed, would be preferable. And why not bore into one of the polar caps, which preserve a record of the palaeoclimates of Mars? Not just one, but many samples will be required, and as many space missions as are needed to deliver them.

Of one thing we can be certain: the sampling missions planned for the years after 2015 will be complex and very demanding. Several launches will be needed, and each mission could cost more than 2 billion Euros. It would not be easy for NASA, which initiated the programme, to finance it alone, especially if it continues to propose the return of American astronauts to the Moon before 2020. With strong backing from the scientific community, Europe could one day play a major part, and even lead, in this new venture. Everything will depend upon the political will of European governments to explore space, and whether they will prolong the ESA Aurora programme beyond ExoMars (see pages 208–210).

12 Astronauts on Mars

From the grandiose projects of the past to today's more realistic programmes

After the Moon... next stop, Mars? The red planet is the only solar system destination to which we might nowadays consider sending astronauts: the surface of Venus, our other neighbour in space, is a lethal furnace. However, many problems remain to be overcome. First of all there is the question of carrying the fuel needed for the return journey. In fact, a journey from the Earth to Mars would be scarcely more expensive in terms of fuel than a journey to the Moon, especially as the Martian atmosphere can be used to slow the descent of the spacecraft, which is not the case on the Moon. However, returning to Earth from the Martian surface, where gravity is stronger than on the Moon, would require extra fuel. Also, there is the problematical question of the time taken for the journey to Mars: using traditional methods of propulsion, each interplanetary voyage would take at least six months. To this must be added time spent on Mars itself. Having landed, astronauts will have to await the next 'launch window', which depends on the relative positions of Mars and the Earth, in order to begin their return journey. In most scenarios, this means a stay on Mars of around 500 days. Briefly, a return trip to Mars would take more than two years. So the vehicles which will make such journeys will have to be robust and have facilities to recycle water and oxygen needed by the crew. Also, crew members will be exposed to high levels of cosmic radiation and the solar wind, with the attendant risk of cancers, and the absence of gravity will affect their physical capabilities as the months go by. For these reasons, early projects were grandiose in scale. In 1952, Wernher von Braun, 'father' of the Apollo programme, envisaged an armada of vessels, each weighing tens of thousands of tonnes. Later, in 1989, President George Bush senior committed his country to a vast project involving the construction in Earth orbit of enormous space ships of more than a thousand tonnes, to be propelled by revolutionary technologies. The estimated cost would have been at least 500 billion dollars! Fortunately, engineers have since developed much more subtle and realistic ideas, and NASA has adopted these concepts for its 'Reference Mission' (see diagrams). With these new concepts, sending astronauts to Mars becomes technologically feasible. Although the project will be extremely costly, it will remain within NASA's budget, especially if other countries contribute to it. As with the lunar programme in the 1960s, scientific research cannot by itself justify such expense. In spite of everything, going to Mars would offer undreamed-of possibilities to

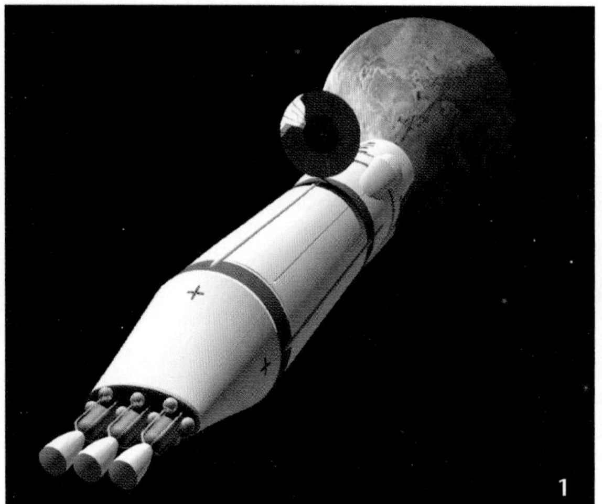

NASA's Mars effort was inspired by American aerospace engineer Robert Zubrin's 'Mars Direct' proposal (1992). This was to use launch vehicles of the Saturn V class, similar to those which took the Apollo craft to the Moon. A key concept is the sending in advance of the astronauts' return vehicle, which will arrive on Mars with empty fuel tanks. These will be filled with 'ergol' (oxygen-methane propellant) in the space of a few months by an automated integral 'mini-chemical plant', powered by a small nuclear reactor, and able to transform atmospheric carbon dioxide into fuel. Only when the vehicle is ready will a craft containing 4-6 astronauts be launched from Earth.

NASA's 'reference mission' strategy predicts four launches for each mission: (1 and 2) the Earth-Mars transfer vehicle, carrying a surface laboratory; (3) a launching assembly/fuel plant, capable of putting a vehicle into Mars orbit; (4) a Mars-Earth vehicle waiting in orbit. Each mission will be able to use equipment left by previous missions. (All courtesy NASA.)

explore, study and understand another world. Will people walk on Mars in the decades to come? In January 2004, President George W. Bush officially committed NASA to this long-term goal, although his first objective was not Mars, but the development of bases on the Moon from 2020. This first stage requires the development of heavy launch vehicles and life support systems such as might be needed for a Mars mission, but at the risk of postponing even further our first footsteps on Mars.

Epilogue

A new road on Mars, travelled by the rover Spirit in November 2004. (Courtesy NASA, JPL-Caltech, Cornell.)

As the first edition of this book was nearing completion, one of the authors received a gift of another book: *Physique de la Planète Mars*, written in 1951 by Gérard de Vaucouleurs, one of the most famous astronomers of the twentieth century, and a great expert on the red planet, and on galaxies. The book was published in English in 1954 with the title *Physics of the Planet Mars*. In his book,

Sunset over the rim of Gusev Crater, photographed by the rover Opportunity. On Mars, atmospheric dust surrounds the white Sun with a bluish halo. (Courtesy NASA, JPL-Caltech, Texas A&M, Cornell.)

de Vaucouleurs presented a rigorous overview of all the available observations of Mars, drawing upon the best current knowledge of planetary physics in order to '*summarise those things which are already known through serious consideration*'. Fifty-two years after its publication, the book made edifying reading: because nearly everything in it was wrong! We read, for example, that '*the permanent atmosphere of Mars can be constituted only of nitrogen, and a small quantity of argon*', though we now know that the Martian atmosphere is essentially carbon dioxide. According to de Vaucouleurs, '*...the surface pressure on Mars is 85* $\pm$ 8 millibars' – the value in reality is 6 millibars. Again, '*we can now confirm that the polar caps are like those on Earth, made of frozen water*', and the author adds in a footnote '*we shall not linger here over a detailed discussion of the reasons why the hypothesis involving frozen carbon dioxide is no longer tenable*' – on the contrary, the caps do indeed contain considerable amounts of carbon dioxide snow.

A message to Earth? This heart-shaped depression on the flank of the volcano Alba Patera was photographed by the MOC camera, on Mars Global Surveyor, in 1998. The feature, 2.3 km wide, was created by stretching of the Martian crust within a trench of the 'graben' type (see pages 100–102). (Courtesy NASA, JPL, Malin Space Science Systems.)

Far from deriding this famous author, we realise that the traps laid by Mars at the feet of the best scientists of the past teach us caution today. We now have direct measurements taken on the Martian surface, and the finest remote-sensing instruments are mapping the planet from every angle. However, contrary to what the title of the present work might suggest, and although we have done our best to draw together current knowledge, we are far from being able to tell the real story of Mars. The only thing that is certain is that our book too will be subject to debate: less than three years passed between the first edition, and the writing of the revised edition now before you – and already, discoveries made by Mars Express and the Mars rovers led us to rewrite whole sections. There is no doubt that, in time, some of the theories that we have laid out in these pages will be contradicted.

Are we pessimistic? Not a bit of it! The aim of this book is none other than to invite you to take part in this great scientific and technological adventure – the exploration of the red planet, and the search for its origins and mechanisms. We hope that we have succeeded in conveying the wonder awaiting us, as Mars continues to offer up more discoveries and surprises.

Index

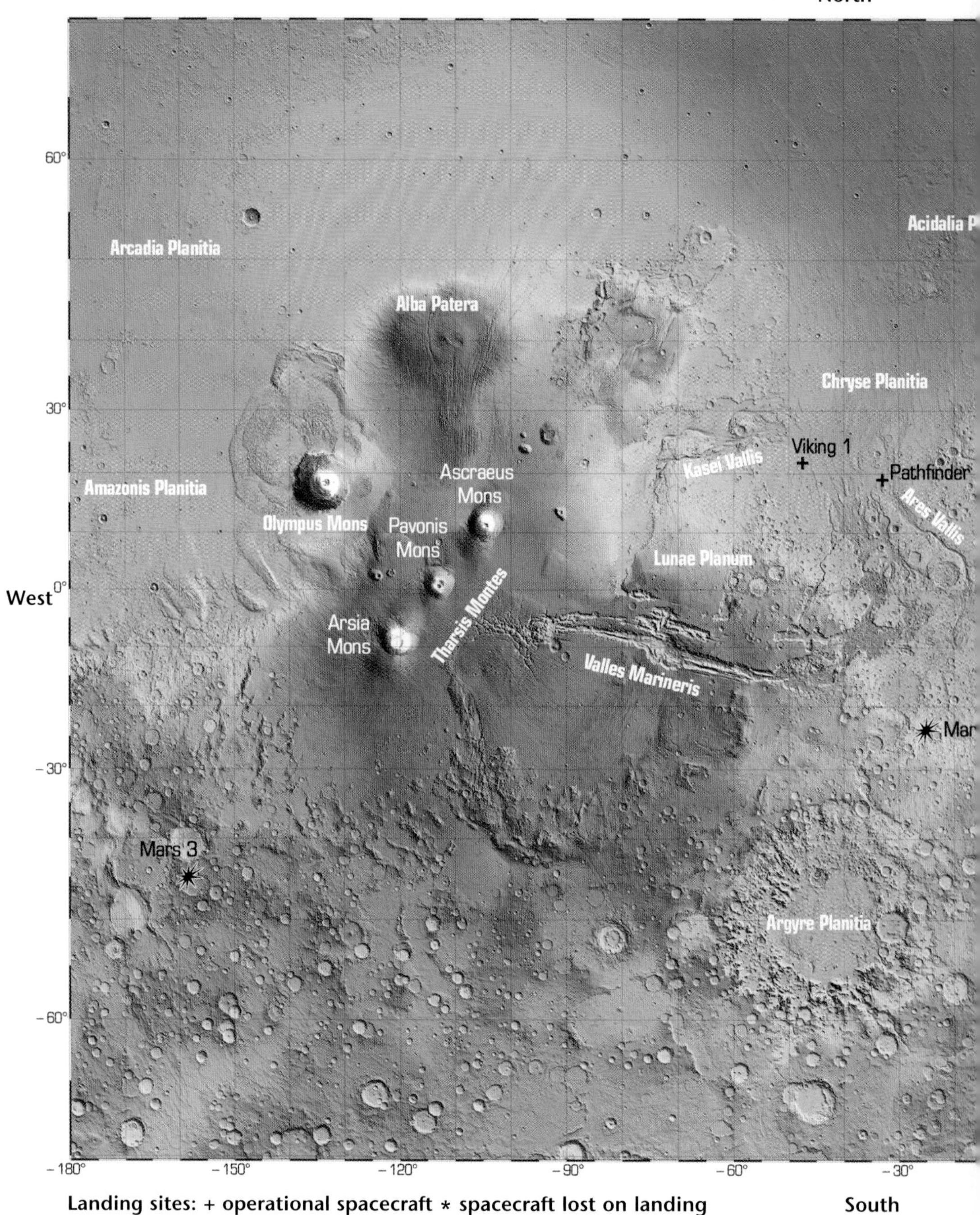

A global topographic map of the planet Mars, obtained from data acquired by the Mars Orbiter Laser Altimeter (MOLA) carried by the Mars Global Surveyor spacecraft. Thanks to this instrument, the global topography of Mars is now known to a greater precision than that of the Earth. The major surface features o Mars, referred to in this book, are labelled here: the volcanoes (Mons, Patera), the plains and impact basin (Planitia), the valleys and canyons (Valles). In the absence of oceans, 'zero elevation' on Mars (the equivalent of 'mean sea level' on our planet) corresponds to the planet's average elevation, or the mean radius of the planet. (Courtesy MOLA Science Team.)

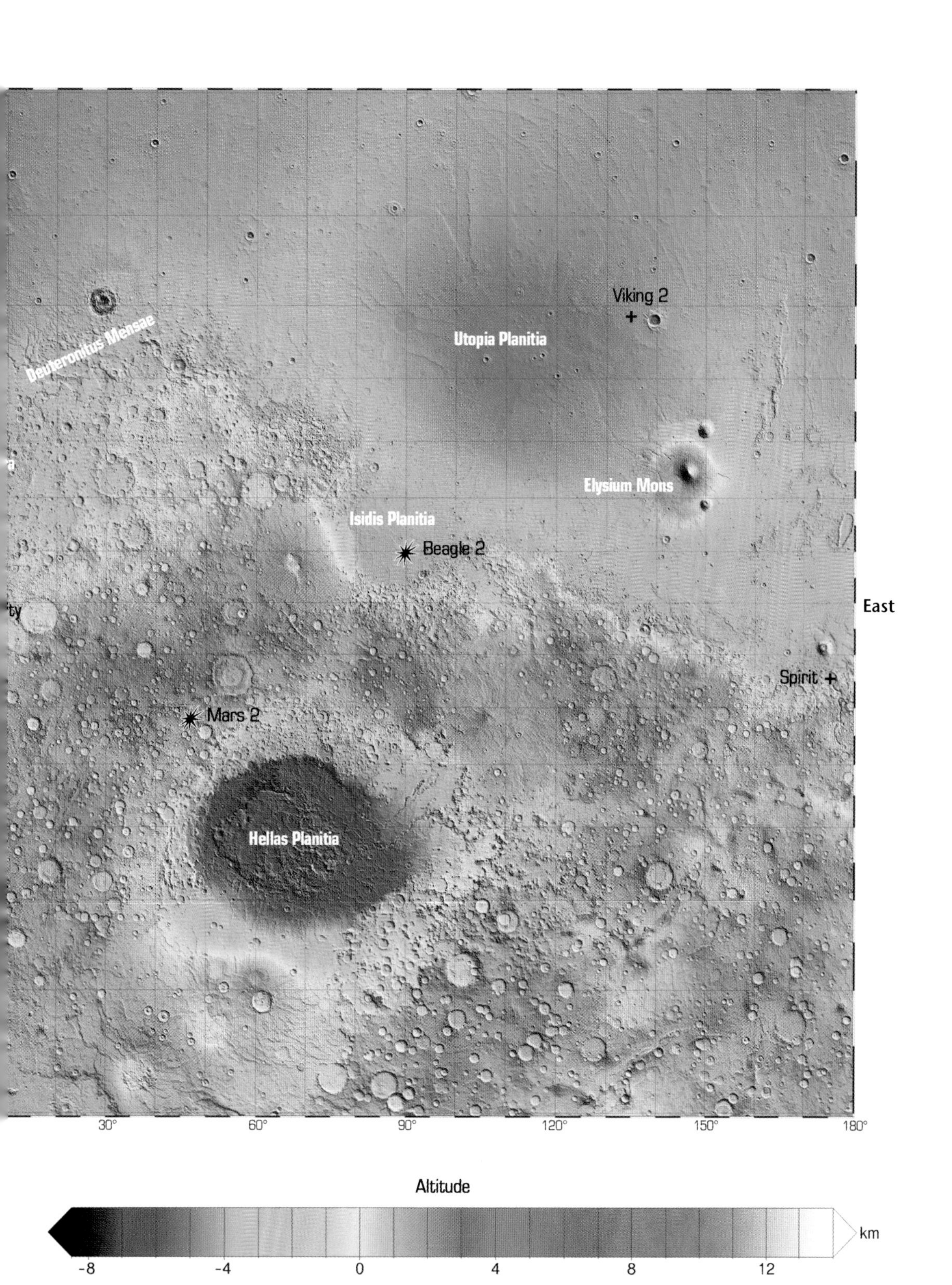

Deuteronitus Mensae
Utopia Planitia
Viking 2
Elysium Mons
Isidis Planitia
Beagle 2
East
Spirit
Mars 2
Hellas Planitia
30°
60°
90°
120°
150°
180°
Altitude
km
-8
-4
0
4
8
12

Printing: Mercedes-Druck, Berlin
Binding: Stein+Lehmann, Berlin